Writer and editor Robert McNally lives in
Chico, California, near the north-south
migratory passage of the gray whale.

So Remorseless a Havoc

So Remorseless a Havoc

Of Dolphins, Whales and Men

BY ROBERT McNALLY

ILLUSTRATIONS BY PIETER AREND FOLKENS

LITTLE, BROWN AND COMPANY · · · BOSTON · TORONTO

FIRST EDITION

All photographs are courtesy of the author. Lines from "Whales
Weep Not" from *The Complete Poems of D. H. Lawrence.*
Copyright © 1964 by Angelo Ravagli and C. M. Weekley,
Executors of the Estate of Frieda Lawrence Ravagli. Reprinted by
permission of Viking Penguin Inc.

Library of Congress Cataloging in Publication Data

McNally, Robert, 1946–
  So remorseless a havoc.

  Bibliography: p.
  Includes index.
  1. Cetacea. 2. Whaling. I. Title.
QL737.C4M36   333.95'9   81-5958
ISBN 0-316-56292-0   AACR2

MV

Designed by Janis Capone

*Published simultaneously in Canada
by Little, Brown & Company (Canada) Limited*

PRINTED IN THE UNITED STATES OF AMERICA

TO ERIC

peace, best and most gentle friend.

But still another inquiry remains; one often agitated by the more recondite Nantucketers. Whether owing to the almost omniscient look-outs at the mast-heads of the whale-ships, now penetrating even through Behring's straits, and into the remotest secret drawers and lockers of the world; and the thousand harpoons and lances darted along all the continental coasts; the moot point is, whether Leviathan can long endure so wide a chase, and so remorseless a havoc; whether he must not at last be exterminated from the waters, and the last whale, like the last man, smoke his last pipe, and then himself evaporate in the final puff.

—HERMAN MELVILLE

ACKNOWLEDGMENTS

◆ ◆ ◆ ◆ ◆

Thanks are, of course, due. Ronn Storro-Patterson and Ted Walker answered questions all along and helped me get to San Ignacio Lagoon. Keith Howell of *Oceans* and Frances Gendlin of *Sierra* let me reuse work already published in their magazines. Einar Vangstein of International Whaling Statistics in Norway was unfailingly prompt whenever I asked for information. And Shirley Manning was good enough to put me in touch with Dick McDonough of Little, Brown.

Terry gave me peace, quiet, and support. Most of all, she gave me love. To her I owe much more than thanks.

CONTENTS

P R O L O G U E

♦　　　　　♦　　　　　♦　　　　　♦　　　　　♦

When the whales surfaced, all twenty-odd of us were scattered about the afterdeck. It had been a hot morning ashore and now it was high noon and even hotter and work enough just to sit in the shade of the tarp and smell lunch cooking in the galley. Then a crewman pointed astern and broke the calm with a shout. *"Ballenas! Dos ballenas!"* he called out. "Whales! Two whales!" I looked to where the man pointed. Two long dark shapes rose from the sea and blew spouts high over the water. The whales were finbacks.

A dozen of us climbed into the skiffs and took off after the whales in hope of a close look. At the approach of the skiffs, the animals separated, one surging ahead as if leading the way, the other hanging back. The first boat followed the leader. The second was nearly upon the trailing whale when it surfaced and blew. The people aboard turned their faces away from the whale's mist and yelled in surprise.

I sat in the third skiff a hundred yards away, with three other passengers. The crewman, who had a round face more Indian than Spanish, stood in the stern and marked the course the blowing whale had taken as it sounded. He guessed where the animal would surface next, and he headed the skiff to his chosen spot. The whale, though, fooled him, coming up a good quarter of a mile away and spouting. The crewman clucked his tongue at himself, again marked the bearing, gunned the outboard. He picked a new spot and brought the boat in slowly, cutting a half-circle. He killed the motor and let the skiff drift with the breeze.

Minutes passed. We spoke little, that little in a whisper. Waves tapped a steady rhythm against the skiff. I checked the exposure on the camera, cocked the shutter, set the focus, just to keep my hands busy. We waited yet. More minutes passed. It struck me all at once how very much alone we were. The first skiff was out of eyeshot somewhere up ahead, still in pursuit of the lead whale, and the second boat had moved off toward the other side of the channel. We were five people in an open boat on the raw sea, and somewhere in the dark water below swam an animal the size of a boxcar. A little chill sent gooseflesh down my spine.

Then the water rolled back, and the whale was surfacing not fifty feet

A finback (*Balaenoptera physalus*) surfacing in the Gulf of California

away and spouting, the sound a rumbling blasting wetness. The black mass of the back rolled slowly past, and the dorsal fin came out as the whale slipped underwater. A slick filled in where the animal had been. Again the whale surfaced, a little farther away this time, an animal of such immensity and such languid passage that it took seeming ages for its length to clear a given point. Then a second slick appeared in line with the remains of the first. The whale had sounded and was gone.

My heart was in my mouth and pounding wildly with excitement. I turned to the others. We started giggling and slapping palms like ballplayers, and the crewman's face beamed roundly. He waited till the thrill had run its course and we quieted, then he headed the skiff back to the big boat. No one spoke. Some unknown spell had been cast and we were not about to break its charm lightly.

This book began with that finback whale in the Gulf of California. I had seen whales before that one, dozens in fact, but always from the insulating haven of a coastal bluff or a good-sized fishing vessel. That finback was different; uninsulated, pure, on its own terms.

As I rolled the memory of the whale over in my mind in the days and weeks following, certain aspects of the experience stood out. I had known before that whales are mammals, but the knowledge meant nothing, just another one of those labeled bits of data tucked away in some dustbin corner of my brain. But when the finback appeared and drew its explosive breath, that long-discarded fact jumped out and became palpable and real. This beast was as fully a mammal, as much an animal to pump warm blood and breathe air, as I myself. The sense of biological commonality was oddly startling.

Neither had the whale risen beside the skiff by accident. The channel spanned five miles of water from island to mainland, and this whale, capable of staying down for fifteen minutes and sprinting away like a torpedo boat, instead chose to surface alongside, as if to satisfy its curiosity. Curiosity implied some sort of intelligence in that massive form.

My curiosity, too, was up. I began investigating whales and dolphins: reading, talking to people reputed to be knowledgeable, watching the animals at every opportunity. The more I learned about whales and dolphins and the more time I spent near them, the more they fascinated me. I was hardly unique in my enthusiasm. These days fascination with whales is widespread, passionate, chic. And that very popularity led me to wonder further. What is this relationship between whales and humans? What does it consist of?

The answer took three parts. First, whales and dolphins excite our curiosity mightily. They appeal to the scientific part of our nature. They are both like us — warm-blooded, air breathing, milk giving — and unlike us — fish-shaped and sea dwelling. Disturbed by the seeming paradox, we try to find out more about them.

Second, whales and dolphins symbolize our working relationship with the natural world. Man the hunter has long hunted whales and dolphins, and for the past few centuries we have treated these animals as natural resources, using them to make products as mundane as lighting oil and lipstick. Now, with many species hunted to the verge of oblivion, we are doubting the wisdom of our former ways. The story of our dealings with the whales and dolphins provides insight into our own wrongheadedness.

Third, and most to the heart of the matter, whales and dolphins touch a chord deep inside us, where emotions are huge and names hard to come by. You can hear people talk of the special intelligence and compassion of whales, of their freedom and peace in being, as if these were new insights. They are not. The chord reaches well back, to folktales and myth, to the Greek and Roman poets who spoke of the innocence of the dolphin, to the Biblical writer who made the whale God's instrument for teaching Jonah a dramatic object lesson, to Herman Melville and his cosmology of fright and beauty bottomed on the whale.

Such is the theme of this book: the threefold source of our fascination with whales and dolphins, "the most devout of all beings."

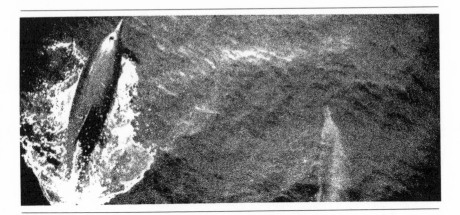

Leviathan's Way

If we choose to let conjecture run wild, then animals, our fellow brethren in pain, disease, suffering, and famine — our slaves in the most laborious work, our companions in our amusements . . . may partake of our origin in one common ancestor. We may all be melted together.
— C H A R L E S   D A R W I N

Science and the Cetaceans

Then, as now, dolphins came to accompany the ships on their way. The Greeks loved to see the animals. They feared the perils of the waters Homer called wine-dark and fish-infested, knew too the paralyzing loneliness of the open, empty sea. Company was welcome, doubly so the company of such graceful creatures coursing and leaping before the prow. No wonder that the Greeks made gods of the dolphins; when the land had dropped away, it was good to have a kindly divinity close at hand.

But the Greek view of the world hardly ended with the ready creation of gods. The Greeks sought to uncover the order of the universe, and that search began with the accurate observation of everything about them. In his several treatises on animals, Aristotle wrote carefully about dolphins. They breathe air, he realized, for they have lungs and when a dolphin is "caught in the nets, he is quickly suffocated for lack of air." He noted as well that dolphins and whales, which he recognized as related forms, bear their young alive and feed them milk produced in "two breasts, not situated high up, but in the neighborhood of the genitals." He called attention to the dolphins' many sounds, their strong maternal attachments, the care they show for hurt companions, their open friendliness toward humans. All in all, there was something distinctly unfishy about dolphins. "These animals are after a fashion land and water animals in one," Aristotle wrote. "For so far as they are inhalers of air they resemble land animals, while they resemble water animals in having no feet and in deriving their food from the sea."

As the conventional wisdom has it, science is progressive, building ever higher on the foundation of what has gone before. Yet the study of whales and porpoises did not advance beyond Aristotle's work for centuries; in fact, it declined precipitously. The Roman Pliny, like Aristotle, wrote on natural history, but unlike Aristotle, he borrowed his information without checking the source. Some of the stories he reported are true and valuable: the dolphin that befriended swimmers on a North African beach, the fishermen and wild dolphins in southern France who worked together to catch mullet. However, Pliny was a literary gossip at heart and given to embellishing for dramatic effect. Great whales so impressed him with their size that he gave the dimensions of one particularly great individual as nearly 1,000 feet long and 500 wide. Pliny also contradicted himself. Sometimes his dolphins breathed air, sometimes water.

Still, in comparison with what followed, Pliny was a lighted window on a dark and moonless night. Between barbarian invasions and Christian otherworldliness, close observation of the natural world became a lost art. Belief supplied the form of the universe, fantasy filled in the detail. In drawings and woodcuts dolphins assumed the guise of fish, complete with scales, long wavy fins, and vertical tails. The great whales were transformed into monsters of the deep. For blowholes they had pipes and chimneys spouting fountains of water; their mouths sported the fangs of dragons; fantastic ruffs, manes, collars, and crests decorated their necks and heads. Canon law considered whales and dolphins fish and allowed their flesh to be eaten on meatless Fridays.

The Renaissance shaped the intellectual life of Europe anew, partly by resurrecting the classical Greek insistence on careful observation. Science, though, requires more than accurate description; it demands as well the arrangement of facts in a way that casts light on the laws and regularities underlying phenomena. Some naturalists set themselves to the formidable task of finding the order in the seemingly infinite variety of living things by arranging organisms in groups based on important similarities. Linnaeus, the greatest of the classifiers, broke with the popular notion that dolphins and whales belonged to the fish. He called them cetaceans and he declared that cetaceans are mammals, a term he himself had coined to refer to all warm-blooded, milk-producing animals.

For the next 150 years, however, the scientific study of cetaceans in Europe advanced little beyond Linnaeus's taxonomic insights. Some dissec-

A medieval woodcut of the whale as monster

tions were performed on dolphins, whose carcasses could be purchased from commercial hunters supplying meat markets, but no similar studies were done on great whales even though several hundred animals were killed each year. This lack resulted in part from the location of the whaling grounds in the distant and dangerous Arctic. It also resulted from a lack of economic pressure to study whales. Although the conventional wisdom again has it that science moves ahead solely in response to its own dictates and not to the pressures of the outside society, the facts are often quite the opposite: microbiology owes its beginnings to the misfortunes of French winemakers and silk growers as well as to the genius of Louis Pasteur. Whales were a natural resource and as long as the supply lasted, little incentive existed to go to the considerable trouble of studying them. What little research there was had a commercial bent. In 1796 the Dutch Academy of Science offered a prize for "the best biological description and natural history of whales, such as would help to discover their habitat and the best methods of killing and catching them." The winning paper included the design for an improved harpoon.

In the heydays of nineteenth-century whaling almost 20,000 voyages were made to all corners of the oceans, yet no more than a handful carried men with a scientific interest in whales. Later, whaling shifted from sailing ships to shore stations and later still to pelagic factories, changes that gave biologists the opportunity to study fresh carcasses on land or deck. Still, systematic studies of whales, even the simple gathering of statistics on whales killed, were slow in coming, and they came finally in the 1920s because of commercial pressures. Concerned that whales were being destroyed too fast to sustain the rich Antarctic bonanza, the British government established an agency responsible in part for research into whales and whaling, mostly whaling. Similar groups in Japan, the United States, the Soviet Union, and other whaling nations began work after the Second World War, as preparations were laid for the second great assault on the treasures of the Antarctic.

The efforts of all these bureaus and committees over the past few decades amount to a heap of reports, papers, and monographs several feet high and several hundred pounds heavy, all of it written in the Latinate jargon of professional science and buttressed liberally with statistics, tables, graphs, and growth curves. Between the sheer mass, the ornate words, and the abundant numbers, you can get the mistaken sense that cetaceans are well understood. Actually the exact opposite is true: less is known of the cetaceans than of any other order of mammals. The animals are difficult to study. Carcasses tell about living whales largely by inference, and the scientist who studies life from the remains of death works with a purblindness as severe as that of the prisoners in Plato's fable of the shadowy cave. Studying living cetaceans, though, is trickier yet. Many species inhabit remote areas where they are hard to find. Even when the whales are located, the would-be observer is limited to seeing surface bits and pieces of animals that spend most of their lives under water. Captivity succeeds only with relatively few cetaceans. Some are simply too large and powerful, others die outright when confined.

So our longstanding ignorance of the cetaceans remains largely intact. Still, in the past three decades particularly, we have come a short way toward understanding the whales and dolphins. Every step along that way has deepened our fascination with these creatures and added new allure to the quest of knowing them fully.

A good measure of the scientific interest in cetaceans comes from their evolutionary oddity. The animals embody the reversal of a historical trend of considerable antiquity. Over countless eons animals had crawled from the water step by step to colonize the land. Then the whales turned around and went back to the water. To be sure, they were not alone. Among the mammals, seals and otters and manatees also went to sea, as did the penguins and auks from the birds. But of all the returnees, the cetaceans adapted most completely to life in the water.

On first meeting, the idea that an animal as aquatic as a whale or dolphin evolved from a land animal seems unbelievable. Superficially cetaceans look like fish, but deeper examination points to the terrestrial origin of the whales and dolphins. The flippers, for example, contain the same bones, albeit much reshaped, as the arm and hand of a human or the foreleg and paw of a dog. Whales lack hindlimbs, but in their body cavities are small vestiges of what once were hipbones and pelvis. Although most cetaceans appear neckless, their spines contain seven cervical vertebrae, the same number found in all mammals from shrews to giraffes. Unborn, embryo whales early in their development look much the same as other mammalian embryos, humans included. For a while, the similarities remain. Rudimentary legs appear, nostrils open at the tip of the snout, genitals develop on the body surface. Then the changes begin, changes similar to those marking the evolution of the whales. Hindlimbs vanish, nostrils move backwards to become the blowholes, genitals disappear inside a slit to smooth and streamline the body surface.

In 1950, Alan Boyden and Douglas Gemeroy of Rutgers University undertook a series of experiments aimed at determining the cetaceans' point of origin among the mammals. Boyden and Gemeroy compared the blood proteins of mammal groups, reasoning that related animals should be more similar in their proteins than distant ones. They found the cetaceans most similar to the artiodactyls — the even-toed grazers like goats, pigs, cattle, and deer.

Boyden and Gemeroy's finding fits with other facts that point the same way. Whales have multiple stomachs, as do the cud-chewing artiodactyls. Likewise, the reproductive system from a cetacean is similar to a goat's or cow's. Cetaceans can even catch the same diseases as domestic stock. Captive dolphins have contracted erysipelas, usually a malady of pigs, and an infection much like blackleg in cattle.

It would be a big mistake, however, to think of the ancestor of the blue whale or the harbor porpoise as some sort of prehistoric billy goat that learned to swim. The animals from which both artiodactyls and cetaceans evolved differed profoundly from the modern representatives of either group.

The period in which the ancestors of the whales turned toward the water was an evolutionary hothouse. The dinosaurs, after some 130 million years as the dominant land animals, were largely extinct. Mammals, originally furtive shrewlike animals active at night when cold slowed the reptiles, evolved rapidly and took the place the dinosaurs vacated. The forerunners of modern horses, rhinoceros, elephants, rodents, and primates appeared. Primitive carnivores preyed upon the various herbivores, which grazed the expanding grasslands. And along the rivers and lakes of this world of 60 million years ago, a group of animals that would become the whales took up their existence.

No fossils of these precetaceans have been uncovered, but paleontologists make the educated guess that the animals resembled the mesonychids, large piglike creatures able to eat both plants and animals. Most likely, the precetaceans lived on land and foraged in water. As time passed, they spent more of their lives away from the shore, returning there only to give birth, as some seals do. Add about 10 million years, and the whales had abandoned the land completely and become wholly aquatic, expanding their range from rivers and lakes to bays and estuaries and from there out into the open sea.

The cetaceans had by now divided into groups, each following its own evolutionary pathway. The archaeocetes, whose name means "ancient whales," are known from the oldest fossils, which date back 50 million years. The early archaeocetes little resembled what we think of as whales. Many types had long serpentine bodies, flippers with elbows, flexible necks, small flukes, even reduced hindlimbs. The head of an archaeocete looked almost crocodilian, with large eyes, a small braincase, prominent teeth, and nostrils about midway between the forehead and the tip of the snout. Later archaeocetes were more familiar; one such animal, *Dorudon osiris,* looked much like a modern dolphin.

The archaeocetes were not the ancestors of contemporary cetaceans, but a line of parallel development that became extinct about 25 million years ago. The other two groups, which once shared the oceans with the archaeocetes, remain today, although they have evolved considerably since then. These groups are the odontocetes, or toothed whales, and the mysticetes, or baleen whales. Odontocetes and mysticetes shared certain aspects of their evolution. The body became increasingly streamlined, and the flukes grew into powerful propulsive organs. The nasal bones migrated backwards till they were flush with the forehead, and the nostrils opened upwards to become blowholes positioned well above the waterline of a whale surfacing to breathe. The movement of the nasal bones was part of an overall recasting of the bones of the skull, which also provided a larger cranium for a brain that was steadily gaining size.

The recasting of the skull differs in odontocetes and mysticetes, and that is but the first of a number of distinctions between them. Toothed whales, unsurprisingly, have teeth, while baleen whales have baleen filter plates instead. Mysticetes have two blowholes; odontocetes, one. The skull of a baleen whale is symmetrical; that is, if you cleave it down the middle, one half is a mirror image of the other. Most toothed whales, though, have asymmetrical skulls. In many species the asymmetry is simply a tilted or offset nasal septum, but in some cases, such as the long spiral tusk of the male narwhal which is actually a single tooth growing from the left upper jaw, that asymmetry is marked indeed. Odontocete males are generally larger than females; mysticete females, by contrast, outgrow their mates. There are even subtle biochemical differences between toothed and baleen whales. Their oils and fats are chemically distinct, as is their myoglobin, the compound that stores oxygen in the muscles.

The oceans, rivers, and lakes of the earth contain about eighty species of cetaceans. Of these, baleen whales account for only ten, but despite their small portion of the whole, mysticetes include most of the animals people think of as whales. The baleen whales are divided into three families, the first of which has only one species, the gray whale. On the basis of its anatomy, its coastal range, and its calving in shallow lagoons, the gray is considered the most primitive of the living mysticetes, an animal basically similar to baleen whales of 20 to 30 million years ago. The other two extant mysticete families are of more recent origin, their earliest members dating back 12 to 15 million years. The balaenid, or right whale, family comprises three contemporary species: the bowhead, or Greenland right whale; the right, or black right, or Biscayan right whale; and the pygmy right whale. Balaenid whales have immense heads with long baleen; the plates reach as much as 14 feet in the bowhead. The balaenopterids, which make up the third mysticete family and include the blue, finback, sei (pronounced "say"), Bryde's, minke, and humpback whales, are, with the exception of the languid humpback, swift and powerful swimmers with seemingly endless stamina. The balaenopterids have up to 100 grooves on their throats that expand like accordion pleats when the whale engulfs its food. From the grooves comes another name for the balaenopterids: rorqual, derived from the Norwegian term meaning "tube fish."

The mysticetes are by and large huge animals. The smallest of the group are the pygmy right whale, which reaches about 20 feet at maturity, and the minke whale, which rarely exceeds 30. The gray, humpback, and Bryde's whales reach lengths in the range of 40 to 55 feet. Seis, rights, and bowheads sometimes make 60 feet. The finback is bigger yet, the females besting 70 feet at full growth. But even the finback is dwarfed by the great blue whale, which is 25 feet long at birth, sexually mature at about 75, and fully grown at 80 feet in the male and 87 in the female. At least a few blue whales have hit 100 feet, and they must have weighed between 150

Narwhal (*Monodon monoceros*), dorsal view (top); flank view (bottom)

and 200 tons. The blue whale is not only the largest cetacean; it is the largest known animal ever to appear on earth.

The toothed whales are smaller than the baleen whales, with only the sperm whale, the killer whale or orca, and two of the beaked whales reaching or exceeding 30 feet. However, odontocetes outnumber mysticetes, totaling some 70 species in all — the "some" arising from scientific uncertainty about the taxonomy of certain varieties. The odontocetes include five species of river dolphin; perhaps twenty kinds of beaked whales, half of which are known from only a few specimens; the sperm whale, or cachalot, and its small relations, the pygmy sperm whale and the dwarf sperm whale; the beluga, or white whale, and the narwhal; and forty to fifty dolphins and porpoises.

The words *dolphin* and *porpoise* cause considerable confusion. To begin with, *dolphin* sometimes refers not to a cetacean but a fish, *Coryphaena*, called *mahi-mahi* in Hawaii. Dolphins, according to many taxonomists, belong to the cetacean family Delphinidae and porpoises to the Phocoaenidae. Unfortunately, the distinctions between these families, however sensible they are to a professional, are less than readily apparent to the lay eye. Dolphins have a prominent beak, a triangular dorsal fin, and peglike teeth. Porpoises lack the beak — as indeed do certain dolphins — have a low dorsal fin — which is entirely absent in the right whale dolphin — and their teeth are flattened and spadelike — except for the Dall's porpoise, whose teeth look like a dolphin's. To thicken this murk further, some dolphins are hardly ever called dolphins — the orca and pilot whale, for example — while fishermen refer to dolphins and porpoises alike as porpoises.

These days scientists, like fishermen, refer to all small cetaceans as *porpoises* and call *dolphins* only those species that belong to the family Delphinidae. This tin-eared practice elevates an undeserving word to generic standing. *Porpoise* derives ultimately from the Latin *porcus-piscis,* meaning pig-fish and referring either to the porpoise's grunting or to the eating quality of its flesh. *Dolphin,* on the other hand, comes from the Greek *delphus,* meaning womb and referring both to the dolphin's womblike shape and to its home in the sea, the origin of all life. *Porpoise* smells of fishmongers; *dolphin* speaks of magic. *Dolphin* is the right and fitting word for the small cetaceans.

In the grand leap of the previous pages from the mesonychids of an antiquity known only from fossils and guesswork to the whales and dolphins of the present day, the specifics of cetacean evolution have been passed over all too quickly. That omission is more essential than it might seem. Evolution is not the elaboration of a grand design; it results instead from the accumulation over time of small, specific changes. The incredible feature of this process of natural selection is that it works so well. When the precetaceans moved into the water, they took with them the physiological and anatomical equipment of the mammals. That equipment had evolved, from bony fishes to amphibians to reptiles to mammals, to adapt to life on the land. The return to marine existence led to a refitting of the precetaceans' land-adapted bodies, and that refitting, the product of innumerable changes and changes of changes over the past 60 million years, yielded the modern cetaceans.

Since water is 700 to 800 times denser than air and that much more resistant to movement through it, swimming in water is a very different proposition from walking or running on land. The cetacean body, originally of the familiar four-legged design, evolved to a shape suited to swimming. The hindlimbs disappeared, the tail developed flukes, and the forelimbs flattened and rigidified into flippers. Yet, despite the seeming fishiness of this adaptation, it retains a distinctly mammalian character. Fish beat their tails from side to side, but whales and dolphins wave their flukes up and down, using the muscles of the back in the same way as a hopping land mammal.

The cetacean spine contains a large number of vertebrae: 93 in the white-sided dolphin compared to 50 in a dog and 26 in a human. The forward portion of the spine is rigid. In right whales, dolphins, and porpoises, several of the neck vertebrae are actually fused together, keeping the head from flopping around and slowing the whale as it swims. Farther down, the many separate vertebrae and the cartilage pads between the

Dolphin fish, porpoise, dolphin

bones make the spine extremely flexible. Along the lumbar spine, which corresponds to the human lower back, lie large muscles attached by long ligaments to the tail vertebrae. The alternate contraction and relaxation of these muscles moves the tail up and down. The tail is flattened from side to side to cut through the water with a minimum of resistance. The flukes flow with the stroke of the tail as if mounted on a pivot, so that water moves backward and the whale forward throughout the up-and-down motion.

The cetacean shape eases the progress of the animal as it swims. The body has few protrusions: external ears disappeared long ago, and the genitals and mammaries are tucked into slits on the underside. The dorsal fin provides considerable stability with little drag, and the flippers work like thin wings for changing depth or direction. Except for facial bristles in the mysticetes, the skin is hairless and smooth.

The microscopic structure of the skin even plays a role in swimming. The cells of the surface skin of dolphins are unusually large, and, instead of being attached to one another, are set singly in a slick mucuslike matrix. These cells slough off easily. British researchers theorize that when the dolphin accelerates, these cells slide away so that the animal actually sheds the very surface causing drag in that first important burst of speed. Captive Dall's porpoises, which are probably the fastest of all the toothed whales and which cannot swim at their usual pace in the confines of an aquarium tank, soon build up an unsightly scum of unsloughed skin that must be sponged off by a handler.

Cetacean skin reduces turbulence as well as friction. Water touching the surface of a passing object tends to adhere, thus flowing more slowly than water farther out and setting up turbulent eddies that hamper the object's passage. In cetaceans, though, the attachment of upper skin layers to lower and of the whole skin to the blubber gives the surface some of the physical characteristics of a liquid. Water touching the whale's surface flows at the same speed as the water farther out, making what is called a laminar flow. The flow over a whale's body isn't perfectly laminar; even in the fastest species some turbulence forms. But overall drag is reduced considerably.

Sometimes the ocean itself presents a demonstration of laminar flow far more dramatic than any human laboratory can. I was sitting in the bow of a charter fishing boat a dozen miles off the coast of Baja California, watching the night and the sea. The water there was bioluminescent, full of microscopic algae that gave off bursts of light when disturbed and that marked with their greenish glow the form of anything passing along the surface. A herd of common dolphins appeared suddenly from the night, rushing pellmell to ride the boat's bow wave. Several dolphins squirted back and forth across the curl, then sprinted ahead to make room for the next group of bowriders. As the sprinting dolphins sped through the calm water ahead, a line of light as thin and straight as the contrail of a high-flying jet followed each animal. I looked back past the stern. The boat left a huge rolling wake that spread out behind in the shape of a giant illuminated fan. The boat needed its big twin diesels just to muscle through its own turbulence, while the dolphins glided along on a path of self-contained least resistance.

The body contour of a particular cetacean roughly indicates how fast that whale can swim. The right whales and the humpback are the bulkiest and the slowest, cruising at two to five knots and maintaining only about ten in a pressured sprint. The sperm whale is thinner and quicker, cruising at five to ten knots and accelerating to twenty in a pinch. Roughly the same figures hold for dolphins, although some species, like the white-sided dolphin, are capable of over twenty knots and the Dall's porpoise can break thirty. The real speedsters in the cetacean order are the rorquals other than the humpback, which loll along at twelve knots, speed-swim for hours at twenty, and hit thirty in a sprint. Sei whales have been clocked at thirty-five knots, about as fast as a high-speed naval patrol boat.

Warm-bloodedness evolved early in the history of mammals and birds as an adaptation to the temperature extremes of the land. Unlike the sea, where local temperatures vary little, the land jumps from hot to cold and back again both daily and seasonally. Since low body temperature makes vertebrate animals sluggish and inactive, reptiles, which do not maintain a constant body temperature, are restricted in their activity whenever and wherever temperatures drop too low and they cannot live in areas of cold climate like high mountains and boreal forests. Much of the success of the birds and mammals comes from their ability to keep a high constant body temperature, which allows them to forage at night and to colonize habitats impossible for reptiles.

The precetaceans took the warm-bloodedness of the land into the water, and today's whales and dolphins maintain body temperatures of about 96° F. The sea has a largely constant temperature in any given region, and even a migratory whale like the finback or gray faces variations of only 45° F from the Arctic to the subtropics, little compared to the 130° F confronting a polar land animal from winter to summer. However, cold presents an unending physiological problem to cetaceans, for the sea, even in the tropics, is colder than the cetacean internal temperature. Adding to this contrast between organism and surroundings, water conducts heat away from the body some twenty times faster than air, which means that, in effect, water is a good deal colder than air. This is the reason why a naked human is comfortable in a room at 75° F but soon chills in a swimming pool at the same temperature. Indeed we lose heat so fast that a human immersed in 32° F water falls unconscious in fifteen minutes and dies soon after. Yet the narwhal, as but one example, lives out its whole life in the Arctic Ocean, where temperatures warm barely past freezing even in midsummer.

One physiological strategy for staying warm is to eat plentifully and well. Whales are prodigious trenchermen. Blue whales consume up to four tons every long day of the Antarctic summer. Bottlenosed dolphins kept at the New York Aquarium each ate some nineteen pounds of herring a day, which works out to a little more than twice as many calories per pound of body weight than a human takes in. The food cetaceans eat is good for producing heat. Protein yields more warmth per pound than other foods, and the fish, squid, and crustaceans on which cetaceans generally subsist provide considerable protein and thus abundant heat.

Another way to keep warm is to keep moving. Working muscles give off heat, and an active animal stays warmer longer than its sedentary counterpart. Cetaceans spend much of their lives on the move, sleeping only in catnaps. Unsurprisingly, whales have a high basal metabolic rate, and the thyroid gland, which secretes the hormones controlling metabolism, is proportionately three times larger than in terrestrial mammals.

A third way of conserving internal warmth is to grow. A large animal of a given shape radiates less of its total heat than does a small animal of the same shape, all because of the geometry of surface and volume. When you blow up a balloon, both the surface area and the volume of the balloon increase. However, the surface, which has only two dimensions, increases more slowly than the volume, which has three. As a result, the inflated balloon has less surface area per unit of volume than the empty balloon. Since radiation from the body surface is a major source of heat loss, a large animal with less surface relative to its volume loses less of its body heat than does a small animal. Thus great whales commonly frequent cold polar seas, while most dolphins inhabit temperate and tropical waters.

A final way to keep heat in is to put a wall of insulation between the warm body and the cold outside. The whale's coat of blubber serves this purpose. The blubber, which lies in a blanket between the skin and the muscles, varies in thickness with the season and the animal's condition, running from about one inch in the dolphins to six or eight inches in sperm and blue whales and actually exceeding two feet in the plumpest bowheads. Blubber is not, as commonly thought, a layer of pure fat with the consistency of warm, overripe Brie. It is instead a complex mass of fibrous, fatty, and connective tissues that feels much like bacon rind. Blubber performs its insulating task well. A scientist working at a shore station put a thermometer into a dead and disemboweled finback when it was first brought in and then again twenty-eight hours later and found that the deep muscle temperature had dropped only 1.5° F. In fact, the effectiveness of blubber as an insulator proved troublesome to the Antarctic whalers, who discovered that they had to process their kills quickly or lose them and the profits they would bring. An intact ungutted whale carcass warmed and swelled with the heat and gases of its own decomposition. Twenty hours after death the meat began to bake, and it fell off the bones when the whale was butchered. At thirty-three hours the oil was no longer edible. If the carcass was left for two days, the pressure of hot gases inside built until, in grisly testimony to blubber's insulating qualities, the carcass exploded and showered steaming, rotten entrails for hundreds of yards.

The flip side to keeping warm is cooling off, and under certain circumstances, such as chasing fast-swimming prey in a tropical sea, the cetacean has to release heat, not retain it. When such a need arises, the flippers and flukes act like radiators. Lacking an insulating coat of blubber, the flippers and flukes contain networks of blood vessels just under the skin. Arteries bring warm blood into these superficial vessels, which radiate the heat into the water. The cooled blood returns to the body through veins that entwine the arteries, partially cooling the blood bound for the flippers and flukes before it even gets there. This system depends on the ability of water to conduct heat, for a cetacean left high and dry quickly overheats. Captive dolphins transported dry have been known to heat up to 108.5° F before dying. Much the same fate befalls stranded whales. Out of the water and in the direct sun, they soon die from overheating.

Blubber serves functions besides insulation. Since fat floats, the blubber coat adds to the whale's buoyancy. Many whales float after death but sink promptly once the blubber is flensed off. Blubber also serves as a food cache. In times of feast, whales build up fat and oil reserves not only in the blubber but also in the bones, muscles, and body cavity, on which they draw in times of famine. Some of the mysticetes feed heavily in the abundant polar summer, then winter in warmer temperate and tropical waters and live off their stored fat.

Early students of the whales, contemplating their anatomy, felt certain that all were creatures of the ocean surface. There was no way, the scientists felt certain, that the mammalian body could withstand the stress of dives below a couple of hundred feet. Then, one day in 1932, the submarine cable off the coast of Ecuador went inexplicably dead. Dispatched to find and fix the problem, the cable ship *All America* winched up the line and found a dead sperm whale caught in it. A loop of the cable had snagged the whale by the jaw, and the animal's struggles to get free had served only to entangle it further and hasten its death. The remarkable thing about this collision was that the whale had hit the cable at a depth of 3,000 feet, far deeper than science thought whales could go. Nor was this unfortunate whale unique. In the years since then, over a dozen dead sperm whales have been taken from cables below 3,000 feet, the deepest at 3,720. Science stood corrected by fact.

The sperm whale and the bottlenosed whale, both predators of deepwater squid, have the deepest range of the cetaceans. A killer whale has been taken from a cable at 3,316 feet; trained pilot whales will dive to 2,000 feet. Tuffy, a bottlenosed dolphin in the navy's marine mammals program, obligingly dove to 990 feet and took his own photograph. Another navy-trained dolphin has dived to 1,700 feet. Baleen whales, which feed mostly on upper-level organisms, do not go as deep as the toothed whales. A cable-entangled humpback was found at 390 feet, probably near the limit for that species, the rights, and the gray. The other rorquals usually dive no deeper than 600 or 700 feet, but there is the case of a harpooned finback that slammed into the bottom at 1,650 feet hard enough to break its neck. Apparently the whale was still on its way down.

Even with the best technology can offer, no human diver can come anywhere near the whales. Scuba-diving below 100 feet is possible but risky.

Commercial hardhat divers can work at 1,000 feet, but the pressure, the cold, and the dark make such depths terribly hazardous.

Reaching the depths a whale does is no mean feat. The animal must be able to withstand the pressure that builds fearsomely as it goes down; it can draw only on the oxygen it carries in its body; the bends are an ever-present and lethal threat.

As you go deeper and deeper down into the ocean, pressure mounts to an all-crushing grip. At the surface the pressure is 15 pounds to the square inch. At 600 feet, it has increased twenty times. At 3,700 feet, the level reached by the deepest of the snared sperm whales, the pressure exceeds 1,800 pounds to the square inch, more than enough to make an unprotected human look like something left behind by a steamroller.

The cetacean body, unlike the human, has adapted to take such pressure. Under pressure, the liplike valves closing the blowholes seal even tighter than they do on the surface, thus keeping water out of the nasal passages and the lungs. The white envelope of the eye, the sclera, is thick and resilient, able to take substantial force without deforming. A thick oily foam that feels and looks much like beaten egg white and a system of sinuses protect the internal ear and allow it to work even at vastly increased pressures. The chest, rather than standing up to the crush, simply yields harmlessly. Many of the ribs are floating or mobile, and they give in with the pressure. In the picture Tuffy the dolphin took of himself at almost 1,000 feet, his chest is crumpled and sunken. The lungs too collapse, the air in them squeezing up into the nasal passages and sinuses.

The whale must be able to hold its breath all the while it is down, and the length of time a whale can go without breathing corresponds to the depth it can reach. The longest Tuffy stayed under was just less than five minutes. The limit for rorquals is fifteen to twenty minutes. Sperm whales are known to dive for up to seventy-five minutes, and there are reports, which may or may not be exaggerated, of harpooned bottlenosed whales remaining submerged for two hours in a vain attempt to escape.

The structure of the lung provides little clue to how whales hold their breath for such extended periods. Proportionately, cetaceans have lungs no larger than do land mammals, and the lungs of the sperm and bottlenosed whales, the deepest diving species, are only half the relative size of a human's. The lungs of a whale are designed not so much to hold large amounts of air against the oxygen needs of a long dive as to fill and empty quickly and completely. The passageways to and from the lungs are short and wide, and the structures called alveoli, where the blood picks up oxygen and releases carbon dioxide, are large and spongy. Because of this structure whales can empty and fill the lungs by as much as 90 percent of their capacity in one breath — compared to 20 percent in a human — and do it in less than two seconds — compared to four in a human. When the whale or dolphin breaks the surface to breathe, it first exhales. The exhalation causes the spout, or blow, which owes its visibility to the atomization of water clinging to the edges of the blowholes, to the condensation of water vapor in the exhaled air, and possibly to small amounts of mucus from the airways. As soon as the lungs are emptied, the whale inhales, closes the blowholes, and submerges. The sound of this routine, as prosaically physiological as it may be, is an impossible and unforgettable mixture of exploding wind tunnel and breaching floodgate.

Curiously, small quick-fill lungs are an asset for a deep-diving air-breather like the sperm whale. Reducing the amount of air in the lungs also reduces the chance of the bends, and since the lungs collapse with pressure and empty what air they do contain into the nasal passages, the whale avoids rupturing a full lung and dying from a hemorrhage. Still, the whale must have oxygen while it is below. That oxygen is stored outside the lungs, and the amount necessary is decreased by physiological measures that come into play as the cetacean dives.

Before heading down, a whale or dolphin often lies on the surface and breathes several times in quick succession, much like a competitive swimmer ventilating before the starter's gun. The oxygen the whale takes in with each lungful combines with the blood and is carried throughout the body. Cetacean blood has a great number of very large red corpuscles, allowing it to take up large amounts of oxygen quickly. In the muscles, oxygen leaves the blood and bonds with myoglobin, the iron compound that gives red meat its color. Cetacean muscles contain large amounts of myoglobin; the sperm whale has so much that its meat is the deep purple-black of a bad bruise.

As soon as the whale begins its dive, the workings of the body shift to conserve oxygen and to send it where it is most needed. The heartbeat drops by half. Blood flows only to the brain, vital internal organs, and spinal cord, all of which must have oxygen. Circulation to the skin, muscles, and digestive system ceases; for the time being, they can do without. The restricted blood flow and the lowered heartbeat, known technically as bradycardia, are found not only in whales and other deep-diving air-breathers like Weddell seals and emperor penguins, but also in rudimentary form in land animals, humans among them. Martin J. Nemiroff of the University of Michigan has found nine cases of people who had apparently drowned, only to be revived later without permanent injury. One of these was an eighteen-year-old boy who spent thirty-eight minutes beneath

Lake Michigan, yet recovered without brain damage. A primitive form of bradycardia saved his life and his mind.

During the dive the whale's muscles first use the oxygen stored in combination with the myoglobin. Once that supply runs out, the muscle cells begin anaerobic respiration, a series of energy-releasing reactions that do not require oxygen. All the while, waste carbon dioxide and lactic acid are accumulating. In humans, a concentration of carbon dioxide beyond a certain limit triggers an automatic response that forces one to inhale — the reason why a petulant child can't make good on a threat of suicide by holding its breath. Whales lack this reflex. Dolphins, and possibly all the cetaceans, have an exactly opposite response: under water they cannot inhale.

When the diving whale surfaces, the heart returns to its normal rate and blood flow into the muscles resumes, bearing oxygen to recharge the depleted myoglobin and to oxidize the lactic acid and carrying out carbon dioxide for disposal through the lungs. Typically the animal will lie on the surface or swim slowly about, blowing again and again, much the way a miler pants after a race. Whales, though, have what looks to us as a rather leisurely way of panting. Because of their ability to empty and fill the lungs almost completely with each breath, they breathe less often than we do. A resting dolphin takes two or three breaths a minute, compared to between fourteen and eighteen in a human, but the rate of respiration accelerates to about six a minute after a dive. The longer the dive the longer the whale breathes on the surface. The Yankee whalers had an accurate rule of thumb that a 60-foot bull sperm whale down for 60 minutes would blow 60 times before returning to the depths.

That rest on the surface may play a part in preventing the bends, a condition that tortures, cripples, and kills scuba divers. Under the increased pressure of a dive, a greater than normal amount of nitrogen from the air the diver breathes dissolves in the blood and other tissues, sometimes until the gas saturates the body. When the diver heads up toward the surface, the nitrogen begins coming out of solution as the pressure drops. Provided that the diver rises slowly enough for his lungs to expel the gaseous nitrogen as it forms, no problem arises, but if he comes up too fast and pressure drops suddenly, the nitrogen forms bubbles, much the way a bottle of soda pop or beer fizzes when opened. The nitrogen bubbles block blood vessels and compress nerves, cutting off circulation and causing paralysis. This is the bends; it is always agonizing and often fatal.

The bends pose a potential threat to any deep-diving, air-breathing animal, but the workings of the cetacean body limit the risk considerably. To begin with, the whale dives only with the air in its lungs, not with a pair of tanks strapped to its back, and the pressure of the dive collapses the lungs and forces what air they contain up into the nasal passages, where little gas is absorbed into the blood. As a result, only small amounts of additional nitrogen enter the whale's body during the dive. Skindivers, who like whales are limited to the air supplies in their bodies, do not catch the bends from a single dive, but they can develop the condition if they dive repeatedly without allowing enough time between descents for the lungs to dump the accumulating nitrogen. Thus the whale that rests and blows after a dive is ridding itself of nitrogen as well as making up for lost oxygen.

Bends may affect cetaceans occasionally. The most dangerous dives a cetacean makes are not to the greatest depths but to intermediate levels. The reason is that a deep dive collapses the lungs completely and keeps nitrogen from entering the bloodstream. At shallower levels the lungs collapse only partially and air continues to diffuse into the body. Theoretically a whale can get the bends by coming up too quickly from a middling dive. This hypothesis may explain one of the many enigmatic observations of cetaceans. Some years ago, Per Scholander studied the levels diving whales reach by attaching a depth gauge to the harpoons of commercial whalers. One finback, though only slightly wounded by the harpoon, dove to 760 feet, surfaced, and died in a matter of minutes. The whalers Scholander was working with said that they had seen other whales die in the same curious way. It is possible that the wounded whale, in its panic and pain, surfaced too quickly and died from the bends.

The mesonychids were, by the testimony of their fossilized teeth, eaters of both plants and animals. The cetaceans moved away from this omnivorous diet to consume only other animals, and they evolved in different ways to exploit the sea's different food resources.

The mysticetes are filter feeders, straining the water for their food. From each side of a mysticete's palate grow hundreds of baleen plates. Baleen is made of keratin, the same protein found in fingernails, hair, and hooves; it is strong, resilient, and lightweight. The plates overlap, like the slats of a closed venetian blind, and their inner edges are frayed and entwined, producing a sieve or filter.

The precise way a mysticete puts its baleen into action depends upon the species. Right whales, which feed on small free-floating animals, or zooplankton, simply open their vast mouths and plow through the water like bulldozers, periodically stopping to shut their maws and swallow whatever the baleen has trapped.

The rorquals feed with a bit more delicacy. Their mouths are smaller and their baleen shorter than the right whales', but their grooved throats can expand like the bellows of an accordion. The rorqual engulfs its prey — either zooplankton or small schooling fish like herring — its throat bulging with water, then it squeezes the water out through the baleen filter and swallows down the zooplankton or fish left behind.

Rorquals may herd their food closer together to make the engulfing routine more efficient. Finbacks, as one example, pack herring into a tight school by swimming round and round the fish in smaller and smaller circles. Humpbacks actually weave a net of bubbles. The whale dives under the feed and exhales a line of bubbles from its blowhole. The whale outswims the feed and rises toward the surface. Blocked to the sides by the bubbles, from below by the whale's body, and from above by the surface, the fish or zooplankton crowd together. The whale then opens its mouth and engulfs the feed. Sometimes two humpbacks work together at bubble-netting, as it is called, with one net of bubbles blown outside the other and the two cooperating whales opening their great mouths to swallow the prey simultaneously.

Gray whales often feed on schooling fish, but they are also the only baleen whale known to sift bottom sediments, mostly for worms and crustaceans. The gray is also the only baleen whale known to have eaten plant matter. A diver in San Diego recently photographed a gray consuming kelp and defecating the floats. It is unlikely, however, that kelp is a part of the gray whale's normal diet. Possibly the animal had an upset gut and, like a dog champing grass, was seeking relief in a cetacean version of an old home remedy.

The toothed whales feed principally on fish and squid, but these animals, like all predators, are opportunists. The sperm whale, for example, is a squid-eater of some note, taking on even the giant *Architeuthis* squid of the deep, yet stomach examinations of killed sperms have turned up various deep-water fishes, ten-foot-long sharks, sponges, spiny lobsters, fishing lines complete with hooks, and an old shoe.

The cetacean whose supposed eating habits have earned it a particularly notorious reputation is the orca. The killer whale has long been depicted as the wolf of the sea, an animal that, like the evil wolf of legend and fairy tale, kills grotesquely and inspires in its victims a panic so crippling that even great whales roll over and offer their still-living lips and tongue to the killers. The orca is also a whale of a glutton. An oft-repeated example of the killer whale's prodigious appetite comes from the dissection of a stranded orca by the nineteenth-century Danish cetologist D. F. Eschricht, who supposedly found in the gut the carcasses of thirteen dolphins and fourteen seals plus a half-swallowed seal in the throat.

This story arises from a misreading of Eschricht's original report. Orcas swallow their kills whole or in large pieces, and hard parts like claws, bones, and teeth, which digest slowly, accumulate in the first of the whale's three stomachs. Eschricht removed these remains, separated them, and decided that they evidenced twenty-seven separate kills. He did not find whole carcasses, nor did he state the period in which the seals and porpoises were consumed. Ian MacAskie, a biologist studying orcas in British Columbia, recently took four hundred seal claws from the gut of a dead orca. He figures the whale actually ate twenty seals the week before it died. "But," he said, "when I publish this in years to come, somebody's going to say I found four hundred seals."

The Eschricht misexample and MacAskie's fears point out the way most stories about the killer whale get started: a small dose of fact compounded with a full measure of fancy. Orcas do kill warm-blooded animals, but they are not the exclusive mammal-murderers they are often made out to be. Studies by Russian and Japanese cetologists show that orcas in polar seas feed largely on small cetaceans and seals, but in temperate waters eat mostly fish.

However, orcas do, at least on occasion, attack great whales. One such occasion was witnessed and photographed by the crew of the research vessel *Sea World* off the tip of Baja California. The orcas, about 30 of them, attacked a 60-foot blue whale as a team, with some hedging the whale in, another attempting to push it under the surface, and the dominant bulls tearing at its body. In the five hours and twenty miles the *Sea World* followed the chase, the blue whale's dorsal fin was chewed off, its flukes shredded, and large chunks of flesh bitten out, including one massive wound more than six feet square. Finally the orcas broke off their pursuit. The blue continued swimming until nightfall, when the ship lost contact with the whale, but almost certainly it died of its injuries.

The blue was young and possibly diseased. Orcas don't usually do so well against healthy adult whales. All the recently observed encounters between orcas and gray whales, for example, have been standoffs, with the orcas finally giving up and swimming off.

From a purely anthropocentric point of view the most pressing question about the orca diet is whether it includes people. Orcas have long been portrayed as man-eaters. A recent example is the novel *Orca,* in which author Arthur Herzog abuses his poetic license by adding a human cadaver to the porpoises and seals in Eschricht's whale. Human remains have never been taken from an orca, nor is there any evidence to corroborate the story told in the book of a Canadian logger killed by vengeful orcas.

Only two verified attacks on humans by orcas, both in the Antarctic, are known: the first on Herbert Ponting in 1911, the second on Morton Beebe during the navy's Operation Deepfreeze many years later. In both instances the orcas came up underneath ice on which the man was standing, in an apparent, and fortunately unsuccessful, attempt to toss him into the water. Orcas hunt seals and penguins in this very way, and it seems likely that the whales, looking up through the distortion of several feet of ice, mistook the man for their usual prey.

Usually it is man who does the mistaking. In the past thirty years, newspapers have reported two alleged orca attacks on the California coast, one in 1952 and the other in 1972. Both reports grew from mistaken identity. The incidents did indeed occur, but the culprits were large sharks, not orcas, misidentified as whales in the heat and panic of the assault. The conclusive evidence in both cases was bite marks that showed the characteristic cutting dentition of a shark and could not have been made by the conical crushing teeth of an orca.

Orcas rarely if ever attack humans knowingly. They deserve healthy respect, not fear and loathing and irrational hatred.

The senses are windows onto the environment, the pathways through which information from outside enters the brain. The world an animal knows is the space its senses create.

The precetaceans took typical mammalian senses into the water, but those organs, adapted over eons to conditions on land, changed in accord with the requirements of aquatic life. Many mammals have a highly refined sense of smell. In whales, though, that sense is of little importance. Baleen whales have a small patch of smell receptors in the nasal passages; toothed whales lack even these vestigial sensors. Taste is likewise unsophisticated, as is usually the case with animals that consume their prey whole.

We think of touch as a function of fingers and hands, which cetaceans obviously lack, but they have a highly developed sense of touch nonetheless. Cetaceans like to touch and be touched. Gray whales rub against rocks, anchor chains, and each other, and some even solicit stroking from humans. Captive bottlenosed dolphins flip over and expose their bellies for rubbing much as dogs do. Cetacean cows and their newborn calves touch almost constantly, and mating whales, for all their size and overbearing passion, stroke each other with their flippers in what can be described only as the gentlest of caresses.

Our eyes, like the eyes of most land mammals, work none too well in the water. In air, light entering the eye is bent by the cornea and the lens and focused on the retina, from which nerve impulses are relayed to the brain. In water, however, the focusing of the cornea is lost because that tissue and water have nearly identical refraction indices. As a result, the image reaches the retina unfocused and the eye is farsighted. Another problem with seeing underwater is the lack of available light. At a depth as shallow as thirty feet, 90 percent of the white light at the surface is lost. Below 1,300 feet the world of the sea is pitch black.

To compensate for the lack of focusing by the cornea, the cetacean eye has a nearly spherical lens with great refraction. The eye is adapted as well to low light. It contains a tapetum, the crystallized layer that gives the eyes of cats and dogs their jewellike nightshine and that acts like a mirror to reflect light onto the retina again and again. The retina itself contains great numbers of large rods, the sensory cells suited to detecting small quantities of light. Cones, which provide sharpness and color vision, are also present but nowhere near so abundantly as the rods.

Different species of whale depend on vision to varying degrees. The Gangetic dolphin never uses its eyes. This species is completely blind, its tiny eyes nothing but nonfunctioning vestiges. The other cetaceans can see, but their eyes are relatively smaller than a human's, and the position of those eyes on the sides of the head above the points of the jaw restricts binocular vision and thus eliminates much depth perception. Some species have a good-sized blind spot straight ahead; had the sperm whale a nose on its great bluff face, it couldn't see it. Bottlenosed dolphins, on the other hand, demonstrate sharp eyesight both in and out of the water. In aquarium shows these animals leap from their tank and delicately pluck small fish from the hand of a trainer high above them.

Yet even the best-sighted cetaceans regularly find themselves in situations where vision is of little or no use. Sperm whales chase squid in the absolute blackness half a mile down. Pelagic dolphins cruise on moonless nights so completely dark that one doubts the inevitability of sunrise. Bottlenosed and river dolphins ascend murky backwaters, thread mazes of shoals and bars, and catch fish fleeing at full speed, all without smashing into each other, the bank, or some sunken hazard.

Only in the past twenty years have biologists come to an elementary understanding of how cetaceans "see" without sight. That understanding began, though, not with whales but with a feared and loathed creature of the night: the bat.

Toward the end of the eighteenth century, Lazzaro Spallanzani started wondering how nocturnal animals get about in the dark. Owls, he discovered, must use extremely sensitive eyesight, for when confined to a chamber so tightly sealed that no light can enter, the birds are helpless. Bats, though, are unhampered by such complete darkness, and Spallanzani de-

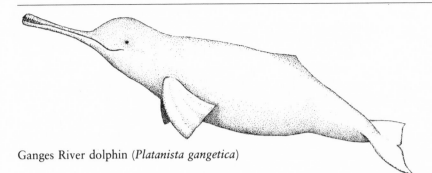

Ganges River dolphin (*Platanista gangetica*)

duced that the animals must be guided by something other than their eyes. To test this idea, he blinded and released four wild bats, then recaptured them several days later. Spallanzani found the bellies of the blind bats as full of insects as those of sighted bats trapped at the same time. Obviously blindness put the bats at no disadvantage in feeding on the wing. Next, Spallanzani plugged his bats' ears. The animals now blundered about stupidly, slamming into walls and obstacles like drunken Lilliputians. Spallanzani tried more experiments but always the results pointed to the same conclusion: as long as bats can hear, they can fly normally.

Unfortunately for Spallanzani, that conclusion was more disturbing than revealing. Bats were known to be as silent as the stones of the caves they haunted, and Spallanzani was at a loss to explain how hearing benefited a mute animal.

The "Spallanzani problem," as it was called, remained unsolved until a serendipitous afternoon at Harvard in 1938. Carrying a cageful of bats, graduate student Donald Griffin walked into a laboratory equipped with a newly developed apparatus for detecting ultrasonic sound. The apparatus happened to be turned on. As Griffin passed it, he noticed that the apparently silent bats were chattering at pitches well above the range of human hearing. Suddenly the Spallanzani problem wasn't a problem; bats have voices after all. Intrigued by his chance discovery, Griffin explored bat sounds further and ultimately showed that bats fly and feed by detecting the echoes of their ultrasonic chirps. This mode of orientation is known as echolocation.

Meanwhile, Arthur McBride, the first curator of what is now Florida's Marineland, was trying to figure out the whys and wherefores of capturing bottlenosed dolphins. His first attempts had gotten nowhere. The dolphins sensed the nets at a distance and fled. Then McBride tried nets with a much larger mesh, and some of the dolphins swam into them unaware and were captured. McBride was a thoughtful observer and he wondered why the one net fooled the dolphins and the other failed. He knew the animals couldn't see either net; visibility in the swamp channels was a foot or two at most. Also the water was stagnant, ruling out any rustling of the nets in a current. Then McBride got a wild idea: what if dolphins echolocated like bats? When the nets were set, masses of air bubbles clung to the shrouds. Such pockets of air reflect sound strongly in water and make pronounced echoes. To an echolocating animal the fine-meshed net looked like a solid wall, while the large mesh gave off only scattered echoes. As McBride saw it, echolocation was the only reasonable way to explain how the dolphins sensed the one net and missed the other.

McBride's insight was ahead of its time, and some years passed before the details of echolocation were worked out. Dependent as we are on light and vision, it is hard for us to understand how an animal can "see" with its hearing. That difficulty is rooted in our limited imagination, however, not in the physics of sound.

Sound comprises waves of high pressure propagated through a medium like air or water or the tines of a tuning fork. The distance from one wave to the next is the sound's wavelength, and the frequency of the sound — that is, the number of waves passing a point in a given period of time — depends on the wavelength. In the air a sound with a wavelength of one meter sends out 344 waves a second. Cut the wavelength to two centimeters and the waves number 17,200 per second. The unit of frequency is the hertz (Hz); one hertz equals one wave per second. Thus the sound with the one-meter wavelength has a frequency of 344 Hz. For the upper reaches of sound, the kilohertz (kHz), equal to 1,000 Hz, is often used to keep the number of zeros in line. Thus the sound with the two-centimeter wavelength has a frequency of 17,200 Hz, or 17.2 kHz.

In common parlance the word for frequency is *pitch*. Low-frequency sounds have low pitch; high-frequency sounds, high pitch. The bottom key of a piano is about 27 Hz, and the top key around 4,200 Hz. Many of the sounds we hear are actually made up of blends of frequency with a peak of energy in one band. Human speech, for example, peaks at about 3,000 Hz.

Frequency depends on velocity as well as on wavelength, and velocity in turn changes with the medium. In air at 75° F the speed of sound is 1,130 feet a second. In water sound hurries along at 4,700 feet per second. This four-times increase in velocity and thus frequency is the reason why sounds change pitch when you are immersed. The outboard motor that chugs gutturally in air changes under water to a berserk buzzing.

For wave energy like light or sound to convey information about the environment, the wave has to be changed by an interaction with that environment. When light strikes an object, part of the spectrum is absorbed and part is reflected. Vision detects the reflected visible light. In essence, echolocation works the same way, by detecting the reflections, or echoes, of emitted sounds.

Low frequencies have far greater range than high. A 100 kHz sound of the loudness of leaves rustling in the wind can travel only 750 yards before dissipating into inaudibility, but a 20 Hz sound of the same intensity loses an insignificant 0.003 percent of its loudness at the same distance. Despite their great range, however, low-frequency sounds have limited use in echolocation. When a sound strikes an object smaller than its own wavelength, it reflects equally in all directions and makes no echo. An echo can come only from a target larger than the wavelength. In water low-frequency sounds have enormous wavelengths. The wave of a 20 Hz noise spans 250 feet, and it can echo only off objects more than 250 feet across, like oil tankers and seamounts. Detecting smaller objects requires the shorter wavelengths of higher frequencies. In various experiments, blindfolded dolphins have repeatedly and accurately echolocated on fish only ¾ inch long, a feat whose physics requires the emission and reception of sounds in the neighborhood of 100 kHz. High frequencies lack range, but they more than make up the deficit in precise resolution.

It is simple enough to understand in outline how a bottlenosed dolphin echolocates. The animal emits a steady stream of high-frequency sound pulses. When a fish swims into range and echoes the pulses back to the dolphin, the animal turns and swims toward the echo's origin. As the dolphin approaches, it sends out pulses faster and faster, often several hundred a second. Judging distance by the interval between pulse and echo and determining the fish's speed and direction by picking up shifts in the echo frequencies, the dolphin closes in and snatches up its prey with the accuracy of a hawk plucking a mouse from the meadow.

That much is easy. The tricky part is the specifics of the process.

Echolocation begins with the emitted sound, and there too the first difficulty arises. Scientists agree that implosive movements of air in the airways create the sounds, but the exact mechanism is a subject of debate.

The sound-making system of the bottlenosed dolphin has been studied the most closely, and John Lilly has done most of the studying. He maintains that the dolphin has three separate sound makers. Two of these are in the nasal passages just under the blowhole. In each passage a muscular tongue comes up against the edge of a membrane, forming a slit much like our own vocal cords. An air sac lies above the tongue-membrane slit, and another sac below it, and the dolphin can make the wide-frequency sounds called clicks and the narrower-frequency whistles by passing air from one sac to another through the slit and back again. Each sound maker works independently. The dolphin can click on one side and whistle on the other. It can also direct the sound as it wishes, probably through the bulging "melon" on its forehead, a deposit of fat that focuses sound much the way a lens bends light.

The third sound maker, according to Lilly, lies in the larynx. That may seem unsurprising, since our own voices originate in our larynxes, but the larynx of the dolphin, as of all cetaceans, lacks vocal cords. Instead, the dolphin larynx has two flat cartilages that can be pressed together to form a slit. When air passes through the mucus-coated slit, a bubble forms. The breaking of the bubble sends a sound traveling forward through the jaws and resonates the teeth, which direct the sound outward in a narrow beam as if they were antennas.

Lilly believes that dolphins use all their sounds — clicks, whistles, and larynx pulses — to echolocate. The combination gives them a vast range of frequencies (about 100 Hz to 160,000 Hz), loudness, blending, and timing and provides them with an acoustic picture of the world every bit as detailed and accurate as our visual impression.

Of all the odontocetes, the sperm whale has the most bizarre way of making and beaming sounds. The anatomy of this animal's huge square head has long mystified whaler and scientist alike, whose many explanations of its structure run the gamut from battering ram to ballast tank. Much of the bulk of the head is taken up by a ligament-sheathed pocket of wax that the Yankees called the case. Early whalers cutting into this reservoir of sticky, white wax thought it mistakenly, and imaginatively, the store of male reproductive fluid and they called the material spermaceti, a politely Greek word for whale semen. The front of the spermaceti case touches a pair of huge lips called the *museau du singe,* or monkey's muzzle, which is located between the air sacs of the right nasal passage. The rear of the case almost reaches the front of the dish-shaped cranium. An air space separates the case from the skull surface, and this space is maintained, even at the tremendous pressures of the whale's deepest dives, by fibrous knobs on the cranium that keep the case from pressing flat against the bone and squeezing the air out. Acoustically this anatomy amounts to two sound reflectors — the air sacs in front and the air space between case and cranium — joined by a superconductor — the case.

An echolocating sperm whale emits one to fifty 5 kHz clicks a second. Each click is actually a series of sounds initiated by a loud blast and trailed by fading repetitions of the opening sound:

CLICK→CLICk→CLIck→CLick→Click→click→clic.→cli..→cl...→c....→

The initial burst comes from the lips of the muzzle banging together.

The sound travels through the case to the skull, where it reflects off the rear air space, rebounds through the spermaceti, hits the front air sacs, and bounces back toward the skull. Each time the sound strikes the front air sacs, some of it escapes to the outside through a ridge of hard tissue that presses up against the skin of the snout to make a sound window. Because some of the echo bleeds off at each reflection, the train of clicks fades away.

Making the sound is only half of echolocation; the whale must also hear the returning echo. Since cetaceans lack external ears, finding the organ of hearing takes a bit of looking. The ear opening in a bottlenosed dolphin is a tiny dimple an inch above and two inches behind the eye, and the ear canal is only $3/16$ inch across. Baleen whales likewise have a small auditory opening behind and above the eye. The ear canal is only $5/16$ inch wide in a finback, and it is blocked by a horny mass called the earplug before it even reaches the internal ear. One wonders how animals so seemingly ill equipped can hear at all, much less navigate and hunt by sound.

The cetacean ear and the human ear contain the same basic components. Sound enters the human ear through the ear canal and vibrates the eardrum. Three tiny bones, the ossicles, transfer the vibration from the eardrum to a membrane in the cochlea. The vibrating membrane sets fluid in the cochlea in motion, and waves in the fluid trigger sensory hairs tuned for specific frequencies. Impulses from the sensory hairs follow the auditory nerve to the brain, where the signals are interpreted.

Under water the human ear works poorly. Trapped air blocks the ear canal and only a portion of the available sound gets through. Also, since water and bone have similar acoustic properties, sound enters the ear through the skull as well as the ear canal, eliminating the distance between the two ears and making it all but impossible to tell where a sound is coming from.

To keep bone-conducted sound out, the cetacean ear is isolated from the skull. Two close-fitting bones, the bulla and the petromastoid, encase the internal ear somewhat like two hands clasped one atop the other. Neither bone touches the skull. The petromastoid hangs on a ligament and the bulla rests on a pad of blubber. The ear cavities are filled with an oily mucus that further insulates the ear and maintains constant pressure inside the organ so that the whale loses none of its hearing when it dives. In its workings the cetacean ear follows the same plan as the human. Sound vibrations move the ossicles, which transfer the movement to a membrane in the cochlea, where waves in the fluid trigger sensory hairs and send nerve impulses to the brain.

The most curious feature of cetacean hearing is the way sound gets from the environment to the ossicles. In baleen whales the eardrum has two parts: a ligament connected at one end to the ossicles and a projection called the glove finger that reaches into the ear canal and touches the hard earplug. According to experiments run by P. E. Purves, sound waves vibrate the earplug and are transferred to the glove finger, which in turn moves the ligament and thus the ossicles. As if to demonstrate once again the deceptive nature of appearances, the ear that looks plugged really isn't.

Toothed whales lack earplugs; they have an even more unusual arrangement. Sound comes to the ear through two acoustic windows that have evolved from — of all things — the lower jaw. The rear portion of the jaw lacks much of the usual bone calcium, leaving a thin blood-and-fat-filled structure that overlies the ear and nearly touches the bulla. Fat deposits extend from the jaw out to the skin and in to the throat. Much as the fatty melon conducts sound out, the fat surrounding the jaw carries sound waves in, where they vibrate the ossicles and set the cochlear fluid in motion.

When a dolphin gives some object the sonic once-over, it listens to the echoes while rotating its head slowly. Various frequencies have differing abilities to penetrate the jawbone. When broad-band high-frequency noises strike the bone at 90°, they go through, but at 9° they bounce off. At various angles between these two extremes some frequencies penetrate and others rebound. By rotating its head and thus changing the angle of the jaws to the echoes, the dolphin can determine the frequencies in the sound and identify whatever it has under acoustic scrutiny.

This system gives the dolphin remarkable discrimination. William Evans trained Doris, a captive bottlenosed dolphin, to earn a reward of fish by selecting a copper plate $1/16$ inch thick. Doris was blindfolded and presented with two targets, one the correct copper and the other any of a variety of test materials. The positions of the targets were changed randomly, so the only way Doris could tell one from the other was by echolocation. Once Doris had the procedure down pat, Evans presented her with glass, plastic, and aluminum substitutes of the same dimension as the copper. She recognized them immediately as counterfeits. Evens then tried an aluminum plate of a thickness calculated to have exactly the same sonic reflectivity as the copper original. Doris still chose the copper. To do so, she had to perceive not only the basic features of the echoes but also their precise frequency composition, an analysis humans can perform only with fancy engineering gear.

The toothed whale's system of hearing, for all its seeming anatomical oddity, is more sophisticated than the baleen whale's. Mysticetes do echolocate, but nowhere near so adeptly as odontocetes. Kenneth Norris, the scientist who figured out how odontocetes hear through their jaws, once strung a line of air-filled aluminum poles across a channel in Scammon's Lagoon to see what would happen. Bottlenosed dolphins detected the barrier at a considerable distance, but a gray whale and her calf crashed through unknowingly, panicked, and, apparently unable to find the first breach, broke through a second time in their rush to escape. Peter Beamish got equivalent results in experiments he ran with a yearling humpback he had rescued from a cod net off Newfoundland. The whale failed to negotiate a maze of rods that would have been child's play to a dolphin. With the possible exception of the blue whale, mysticetes generally vocalize below 2 kHz. These low frequencies allow the whales to determine the topography of the waters they cross and to locate the large masses of fish and krill they feed on, but they permit none of the precision attained by the odontocetes.

For every question science answers about whales, ten more arise. In a way, this endless spawning of inquiry is inevitable. Science never ceases looking about, and given both the difficulty of studying cetaceans and the short history of this branch of biology, it is hardly surprising that questions outnumber answers. But there is more here than an insufficiency of information. The biggest questions may not admit of answers in terms we are used to.

We remain ignorant about much of the cetaceans' basic biology. The function of the retia mirabilia, a spongy network of blood vessels along the spine, is uncertain. Likewise, we do not understand the physiological mechanics of how marine cetaceans maintain the proper salinity of their body fluids while living in an environment saltier than they are, nor has anyone even tried to explain how or why in all the tens of thousands of blue whales butchered in the Antarctic no cancerous tumor was ever found. The ignorance widens when the focus of inquiry shifts to life cycle, demography, and social organization. The cetacean lifespan is still up for grabs, with educated guesses, like a scientific parody of a football pool, claiming every age from twenty to ninety. Population figures for species like the gray and humpback, which migrate close to land, are fairly accurate, but no reliable way exists at present to count the smaller cetaceans of the open ocean. As to our understanding of cetacean society, it is unduly complimentary to call it cursory. Many species live in groups, and that statement nearly summarizes our sociological knowledge of cetaceans.

Yet, of all the questions in the realm of inquiry, those concerning anatomy, pathology, demography, and sociology are the easy ones. They are problems; they admit of solution. The answers require more reliable statistics, better methodologies, additional parameters in the computer models. The really knotty questions are those other ones, the ones we are unsure how to ask.

I have watched whales and dolphins for hours on end and I have watched people watching whales and dolphins, and always the same thing happens: the animals draw the watcher in. Finally the watcher looks at the cetacean and it looks back and something passes between them. And in this curious, unnamed intercourse you begin to wonder. Is there a knowing intelligence behind that glistening eye?

* * * * *

It is a very funny situation. We know the behavior of an electron is not completely determined by the laws of physics. You believe that the behavior of a human being is completely determined by the laws of physics. Electrons are unpredictable, people are predictable. And you call this psychology.

—ARTHUR KOESTLER

The Talking Dolphins

In a way, it is surprising that John Lilly ever came to study dolphins at all. He is a physician by profession, once a professor of medical physics and a student of neurology and psychoanalysis. But while working for the National Institute of Health as chief of the Section on Cortical Integration in the 1950s, Lilly became interested in uncovering the mind within the brain. To get a perspective on the human brain, he wanted to examine a brain of roughly equal size. The closest match was the bottlenosed dolphin.

Marine Studios in Florida offered several interested scientists, Lilly among them, six dolphins as experimental animals. The plan was to map each dolphin's brain by running electrodes into it, a procedure calling for an anesthetic. But general anesthesia, it turned out, actually killed the dolphin before it was anesthetized. Cetaceans breathe consciously, and the anesthetic shut off the conscious brain and the dolphin's breathing. The scientists tried to keep the animals breathing with a mechanical respirator, but attempt after attempt failed. Soon five of the six dolphins were dead. Deciding to spare the last one, a charming but dull bottlenose named Priscilla, the group called off the research, posed for a portrait with Priscilla at her tank, and went home.

In his own lab, Lilly wrestled with the anesthetic problem. In time he developed a way of inserting the electrode into the brain with only a local painkiller. The electrode studies showed Lilly that the dolphin brain has areas of positive and negative motivation, just like the brains of other mammals. When an electrode in the negative area fired, the dolphin gave out a distress whistle. Stimulation of the positive area felt so good to the dolphin that the animal could be trained quickly to stimulate itself by touching a switch with its beak. The remarkable thing was how fast the dolphins learned to stimulate themselves. It takes hundreds, even thousands of trials for a monkey to learn how, but the dolphins picked it up in only twenty attempts.

Fascinated by the work, Lilly determined to study dolphins full-time, and he established research facilities in Miami and the Virgin Islands. Soon Lilly wrote *Man and Dolphin,* in which he indicated that toothed whales have a high-order intelligence and reported that his experimental dolphins had imitated human speech.

Lilly was now pursuing two lines of inquiry. One was the study of communication with the dolphin. The other was the exploration of the deep self in the isolation tank, in which the subject floated motionless for hours and let the mind take its own course. In Lilly's view the two inquiries were linked inextricably; communication with another mind would inform us about our own. The federal government licensed Lilly to experiment with LSD. He took the drug repeatedly and hit upon the idea of combining the drug experience with isolation. Unsure of how the drug would affect a mammal in the water, he first gave LSD to six dolphins. Satisfied with the results, Lilly himself took LSD before a session in the isolation tank.

Lilly's interest in LSD extended beyond gathering experimental information. Lilly had lost his old beliefs and he was seeking, in some pain, after a new Truth. His marriage came apart; his scientific work lost its savor; he nearly died in a medical accident he felt was subconsciously suicidal. For John Lilly, life was an endless dark night of the soul.

All the while, Margaret Howe, an assistant in the Virgin Islands lab, was working further on communication with dolphins. For two and a half months, she spent night and day in a tank with a young bottlenose named Peter and tried, in Lilly's words, "to brainwash him into speaking English." Howe's work with Peter occupied much of Lilly's next book, *The Mind of the Dolphin.* The book delivers far less than its title promises, but it was widely read and reviewed. People began talking about dolphin intelligence and a new tongue called delphinese as if they were proven findings. These notions took on an existence quite apart from whatever Lilly had or had not written.

Lilly, in the meantime, was undergoing a change of scientific heart. Early in the research on dolphins, for example, he had consented to the killing of three animals for brain preparations. Now he abhorred such research. He closed the Virgin Islands lab because, he writes, "I no longer wanted to run a concentration camp for my friends."

Lilly moved from dolphin research into the vanguard of the consciousness movement of the late 1960s. He spent time at Esalen, and he went to Chile to work with Oscar Ichazo. He wrote books about the mind and its doings, and he remarried.

Only recently has John Lilly returned to his work on dolphins. He calls his current research the JANUS (for Joint Analog Understanding System) project, and it involves using a computer as a transformer between human and dolphin to ease communication. Lilly also continues to explore consciousness. He has an isolation facility and he conducts seminars, which often intertwine cetaceans and consciousness. A pamphlet advertising a workshop by John and Antoinetta Lilly says that the topics covered will include "many areas . . . including diet, yoga, body movement, dyadic relationships . . . making planet side trips work . . . working with dolphin interspecies communication . . . etc."

Lilly's work draws fire from the scientific establishment, which holds that his research has nothing to do with real science. The contention is true as far as it goes — Lilly is indeed marching to the beat of a different drummer — but it stops short. Lilly's ideas cut to the heart of how science conducts its business and how humankind views itself. Given their import, these ideas deserve serious consideration and examination.

Lilly argues that humankind has to get itself off dead center in its biological cosmology. "We must strip ourselves . . . of our preconceptions about the relative place of *Homo sapiens* in the scheme of nature. . . . We cannot continue to insist that man is at the top of the evolutionary scale and that no further evolution is possible" (*Lilly on Dolphins,* p. 3). Lilly says that the toothed whales are perhaps our equals: "It is probable that their intelligence is comparable to ours, though in a strange fashion" (*Lilly on Dolphins,* p. 19).

There is purpose to Lilly's supposition of intelligence. He wants to establish communication with dolphins. Like Dr. Doolittle, John Lilly wants to talk to the animals: "Our major goal is breaking the interspecies communication barrier so we can mutually exchange information. . . . Until we live with dolphins and they with us, the dolphin mind remains opaque to us, and the human mind remains dryly out of reach of dolphin teaching" (*Lilly on Dolphins,* p. 200). Such contact will better the sorry state of the human species. "Sometimes I feel that if man could become more involved in some problems of an alien species, he may become less involved with his own egocentric pursuits and deadly competition within his species, and become somehow a better being" (*Lilly on Dolphins*, pp. 207–208).

Lilly's belief in the cetaceans' high intelligence rests on two lines of evidence. The first is the size of the whale brain. We know that our large brains, which average about 1,400 grams in weight, have much to do with our humanness and that the growth in the size of the brain was a central feature of our evolution as a species. Toothed whales have brains as big as ours and bigger. The bottlenosed dolphin has a brain of 1,600 grams, the orca of 4,500, and the bull sperm whale of 9,200 grams — over twenty pounds, the largest brain on earth. The size of the brain roughly indicates the organ's complexity. Generally, a bigger brain is a better brain.

The second line of evidence is the toothed whales' behavior, which Lilly says demonstrates the kind of intelligence expected in such large-brained animals. In *Man and Dolphin* Lilly relates an incident he heard from an unnamed scientist while visiting Norway in 1958. A Norwegian fishing fleet working the Antarctic the previous summer was beset by several thousand killer whales stealing fish from the nets. The fishermen radioed the whaling fleet for help, and several catcher boats with harpoon cannons obligingly steamed over. One of the catchers blasted an orca. Within half an hour all the killer whales over a fifty-square-mile area had disappeared — but only within range of the catcher boats. Fishing craft at a distance from the catchers still had killer whales about them. Clearly, the orcas could distinguish fishing boats from catcher boats, a remarkable distinction, since all of the vessels were converted World War II corvettes and thus identical below the waterline. The only difference between them was the cannon in the bow of the catchers, and this crucial bit of information must have been transmitted by the dying orca to its fellows.

But cetacean behavior isn't just intelligent; it's also ethical. Cetaceans, according to Lilly, are constrained from harming humans. He maintains that, for all their strength, captive cetaceans have never harmed humans. This further shows the intelligence of the whales and dolphins. "The Cetacea realize that man is incredibly dangerous in concert. It is such considerations as these that may give rise to their behavioral ethic that the bodies of men are not to be injured or destroyed, even under extreme provocation. If the whales and dolphins began to injure and kill humans in the water, I am sure that the Cetacea realize that our navies would wipe them out totally . . ." (*Communication between Man and Dolphin*, p. 8).

Intelligence and ethics: such phenomena suggest a mode of communication sophisticated enough to call a language. Lilly maintains that dolphins do send and receive information by sound and, even more marvelous, in captivity they learn human words for talking to their handlers. The dolphin, like the human, is the creature and creator of language.

This, then, is the dolphin — and perhaps the whole kingdom of cetaceans — according to John Lilly. It has great intelligence. It has ethics. It has its own language, and in captivity it tries to speak in the captors' tongue. It is a being altogether very much like ourselves.

John Lilly's interest in the bottlenosed dolphin began with the near-human size of its brain, and brain size has remained the basis of his argument for high intelligence among the cetaceans. Lilly pays little attention to even the gross anatomy of the organ other than to say that it is complex and highly developed — as are the brains of all mammals. It is the size, the sheer mass, of the cetacean brain that impresses Lilly. "The relationship between absolute mammalian brain size and intelligence," Lilly writes, "is yet to be determined. . . . However, we assume that their [cetaceans'] mental operations, within their particular evolution and in their particular environment, may be comparable and even superior to our own" (*Communication between Man and Dolphin,* pp. 246–247).

The cetacean brain has grown so impressively large, as Lilly sees it, because of the physical advantages of life in the water. The brain is a semi-soft mass that can be easily harmed by nothing more than a blow that strikes the head at a glancing angle. If the blow twists the head fast enough, the brain tears free from the skull and shears off the blood vessels entering it. The injury is commonly fatal. The vulnerability of the brain to such injury increases exponentially with the brain's radius. Thus, the bigger the brain, the more subject it is to injury. The human brain is as big as it can safely get. As it is, a fall from a standing height to the ground can tear the brain loose. Any bigger brain must have a much bigger head and skull to protect it. But, because susceptibility to injury increases with size, there is an upper limit to brain size on the land. Once the brain is so big, no skeleton can support the massive head needed to protect it. Aquatic animals, though, need less brain protection because the viscosity and density of water dampen the impact of blows to the head. As a result, aquatic animals can develop much larger brains than terrestrial animals.

This argument runs directly counter to the usual scientific explanation that whales have big brains because their big bodies require equally big central monitoring. The idea makes Lilly laugh. "All you have to do is look at the whale shark to spike that one," he said in a telephone interview. "Forty tons and it's got a brain smaller than a macaque." The standard scientific argument is that brain size is relative to body size. Lilly maintains that absolute brain size is the measure of interest.

The relationship between brain size and intelligence is by no means as tidy and unequivocal as Lilly would have it. Within a single species, absolute brain size means little. The brain in men averages about 1,375 grams, in women about 1,225. Individuals vary widely from this mean, in part because of body size. An oft-cited contrast is between Lord Byron and Anatole France, literary giants both, but Byron's 2,200-gram brain was twice the size of France's 1,100. Einstein's brain was average, and the Neanderthal humans had about 100 grams more than we modern humans do. There is, however, an apparent minimum brain size for a normal adult. People suffering from microcephalus, a developmental disorder in which the bones of the infant skull close too fast and leave the brain no growing room, have adult brains of only 450 to 900 grams and suffer severe mental retardation.

These individual variations, though, actually cover little of the range of brain size in vertebrates, which runs from about 0.01 gram in a goldfish to over 9,000 grams in the sperm whale bull. And we humans are very close to the top of this scale, a fact that helps Lilly's argument immensely. Since we consider ourselves smarter than the others, it does our hearts good to find that we have far bigger brains than the rest of that dull lot. It also helps Lilly's equation of brain size and intelligence that the only animals with brains bigger than the human are the very animals that Lilly wants to argue for, the cetaceans.

One animal, though, untidies this neat happenstance: the elephant. Elephants have brains of some 6,000 grams, four times the size of a human's brain and significantly bigger than an orca's. In his early books, Lilly says nothing of the elephant. Only in his latest work, *Communication between Man and Dolphin,* does Lilly mention the elephant, but then merely in passing. Given Lilly's own insistence on the absolute relationship between brain size and intelligence, the elephant deserves closer attention.

Since most of us see elephants as circus performers and zoo inmates, we tend to write them off as shambling brutes, but elephants are very creative animals. In Thailand elephants in teams of two work in the timber industry moving felled logs to a stream for transport to a mill. When a handler trains a new elephant to the task, he first teaches it specific commands like move right, lift, move left, and so forth. Then he teaches the elephant to move logs by directing its actions with the commands. But once the elephant gets the idea that the point of all this is moving the logs, it does the job on its own, without orders from the handler, who simply supervises from the elephant's broad back. The performance is a long way from rote.

In the dense forest, each log is a new problem, and the two elephants in the team have to size up the situation and figure out a strategy for getting the log to water. The elephants show considerable insight, which they also apply to making mischief. Belled like cats to mark their movements, elephants bent on raiding banana plantations at night have been known to stuff their bells with mud to still them. The elephant has more than an indelible memory; it has a very good set of wits.

Still, studies of wild elephants do not support any argument that the elephant is the undiscovered genius of the planet. For example, E. O. Wilson, author of *Sociobiology,* holds that the elephant's intelligence is comparable to that of the quicker-witted Old World monkeys.

John Lilly to the contrary, body size does make a difference, a big difference. The cells that make up the bodies and brains of vertebrate animals are of roughly equal size. A cell from the gut lining of a goldfish has about the same dimensions and weight as a cell from the gut lining of an elephant. What makes the elephant bigger is not bigger cells but more cells. There is a relationship between the number of body cells and the number of nerve cells needed to monitor and control them. As body size and cell number go up, so too must nerve cell number and brain size.

Harry J. Jerison of UCLA examined the question of brain in proportion to body by plotting the figures for a great number of animals, both living and extinct, and studying the data statistically. The work, published as *Evolution of the Brain and Intelligence,* shows that evolution has involved enlargement of the brain. As one class of animals evolved into another, the brain has gained in size. As a result, fish and reptiles have smaller brains than birds, which in turn have smaller brains than mammals. This finding scotches Lilly's comparison of the whale with the whale shark. Animals from different classes are not comparable. The whale shark's brain makes sense only in shark terms. As for the whale, it must be compared only with other mammals.

Jerison also found that body size does influence brain size, because of the necessary relationship between the number of nerve cells and the number of body cells. However, since the brain increases in size more slowly than the body, at about two-thirds the rate, the precise relationship between brain and body is complex. The probable reason why the brain grows more slowly than the body is the surface-to-volume geometry of increasing size. Much of the brain handles information coming in from the surface, and since surface grows less than volume with increasing size, the brain needs to grow at a slower rate than the body.

To tell just how much brain an animal has in relation to its body, Jerison developed a measure he calls the encephalization quotient, or EQ. If an animal has the average size brain for its body weight, its EQ is 1.0; if the brain is smaller, the EQ is below 1.0; if higher, above 1.0 In everyday terms, the EQ tells us how much more brain an animal has than its body needs — in other words, how much of its brain is "extra."

Humans have the highest EQ of any land animal, and the rest of the primates, our closest evolutionary relatives, follow right behind. But the small toothed whales actually come out higher than the apes and monkeys. The EQ of the bottlenosed dolphin almost exactly equals the human value. The narwhal, whitesided dolphin, and beluga are better endowed than the gorilla, and the orca is brainier than the baboon. The trend to large brains falls off markedly in the great whales. The sperm whale, the humpback, and the blue all have unimpressive EQs below 1.0. Jerison notes, though, that all mammals weighing over a ton, including the elephant, have seemingly small brains. Probably the natural selection that directed the evolutionary increase of the body acted independently of the selection to enlarge the brain, and, in the case of whales, the increase in size involved tissues like blubber that are under relatively little nervous-system control. In other words, for evolutionary reasons all their own, the great whales may have bigger brains than the figures show.

Jerison's work makes two important points. First, Lilly simply cannot get away with his insistence on gross brain weight as an absolute measure of intelligence. Second, though, Jerison shows that small odontocetes have brains in the same major league as the chimp, gorilla, and human.

Brain size means little all by itself. The brain has to be put to some use. The use Lilly focuses on is communication.

Cetaceans communicate in a number of ways. The bright markings on some species, like the killer whale and Dall's porpoise, help the animals recognize their own kind. Cetaceans also communicate with loud noises, slapping their tails on the surface or throwing themselves from the water and landing thunderously. These noises, depending on context, signal anything from alarm to sexual excitement. Whales and dolphins also use a great variety of sounds made in their airways to send each other messages. Some workers, Lilly among them, believe that in the bottlenosed dolphin, and perhaps in similar species, these sounds are complex enough to constitute a language.

However, Lilly has been less concerned with what dolphins have to say to one another than in what they try to say to humans, in human sounds, no less. Lilly first observed this effect early in his work. Playing back tapes of a dolphin making noises during a brain study, Lilly heard what he was sure were sounds uncannily similar to his own dictation into the tape recorder. The dolphin apparently was mimicking him.

Beginning in the summer of 1960, Lilly set to work to study mimicry in a young dolphin named Elvar. For over two years, Elvar was isolated from other dolphins but allowed daily contact with humans. The idea behind this rigorous regimen was to force the dolphin, a social and vocal animal by nature, to direct all its social and vocal behavior to humans, not other dolphins. Vocalization under these circumstances is by no means easy for the dolphin. Dolphins normally vocalize underwater with the blowhole closed. To communicate with a human, the dolphin has to sound off in the air with the blowhole open, and, because humans can hear at all only below 20 kHz and well only from 3 kHz down, the dolphin must drop its voice to the lowest range of its high register to make itself understood. Lilly was trying to get Elvar to accomplish a task every bit as difficult as it would be for us to talk through our noses under water in a basso profundo. Yet, Elvar did just that. He imitated human words and certain simple phrases. When Lilly reported the findings in late 1962 in a paper entitled "Vocal Behavior of the Bottlenosed Dolphin," he and his associates were trying to teach Elvar and two other dolphins, Chee Chee and Sissy, to use words only in proper situations and not at random — that is, to stop parroting and use the words with meaning. The difference is substantial.

Carl Sagan, the astronomer and cosmologist, met Elvar at Lilly's St. Thomas lab in 1963. The dolphin had just returned from a conjugal visit with the females and was in properly high spirits. He approached Sagan, who was wearing a raincoat over his clothes, and rolled onto his back like a dog to get a tummy rub. Sagan obliged him. Elvar swam off across the tank and returned with his belly six inches under the surface. Sagan took off his raincoat, rolled up his sleeves, put the raincoat on again, and rubbed Elvar. Off the dolphin swam, returning belly-up a foot below the surface. Sagan adjusted his clothes once more and again rubbed Elvar, who swam away and came back a full three feet underwater. To reach him, Sagan would have had to undress from the waist up, and that he refused to do. Suddenly Elvar stood on his tail, towering high above Sagan, and said, "More!" Sagan, amazed, rushed out and told Lilly. "Was it in context?" Lilly asked laconically.

Lilly's work on dolphin speech culminated in a paper entitled, majestically, "Reprogramming of the Sonic Output of the Dolphin: Sonic Burst Count Matching." The research involved a human speaking a series of nonsense syllables in a loud voice to a dolphin, which was trained to re-

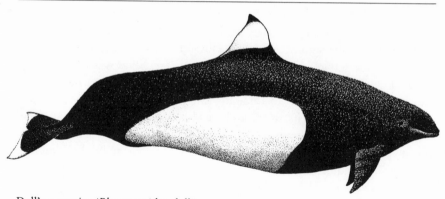

Dall's porpoise (*Phocaenoides dalli*)

peat. The point was to see whether the dolphin could match the sounds, or sonic bursts, in the nonsense series. The dolphin-human exchanges were taped and the accuracy of the dolphin's responses scored. In 200 runs, many of the dolphin responses were perfect; errors were few and far between. In addition, even though the series varied randomly in length, the dolphin never started its repetition before the human finished. To boot, the tapes showed definite mimicry, not just the same sound made over and over the correct number of times. Finally, Lilly made the point that the dolphins performed because they wished to, not because they were rewarded with food. Such symbolic reinforcement, Lilly argued, is unique to dolphins and humans and unknown among such animals as rats, cats, and monkeys.

In fact, the statement that dolphins learn without fish and because of their own "symbolic-reinforcing" programs is nothing more than Lilly's breathless rediscovery of the wheel. Trainers who work with animals that can learn widely and well — a large group including, among others parrots, crows and ravens and jackdaws, sea lions, dolphins, elephants, and dogs — find that once they establish a routine of rewards for learning and the animal picks up the point of the enterprise, they can cease rewarding the animal for each correct performance. The animal learns because it wants to, because learning is fun. Konrad Lorenz, the twentieth century's premier student of animal behavior, says that it comes as a matter of course to researchers in the field that animals enjoy learning. Lilly's claim that the dolphins' performance for the sake of performance was unique to dolphins is baseless. The same phenomenon appears in many animals, many of them distinctly small-brained.

This is hardly the paper's only instance of puffery. Lilly constantly inflates his language — using *biocomputer* for *brain, interlock* for *hookup, reprogram* for *change in behavior, output* for *action* — as if to lend to the phenomena more importance than they themselves will bear. Indeed, what Lilly reports is less significant than his words imply. A dolphin, which is a vocalizer every bit as much as a canary is a singer, is trained to vocalize in the air. Then the human speaks a random number of nonsense syllables, and the dolphin speaks back the same number in a voice that sometimes mimics aspects of the particular human voice and sometimes does not. Interesting, yes, but nowhere near as universe-rattling as Lilly holds it out to be.

Lilly's experiment shows that a dolphin can learn to vocalize in the air — a considerable feat — and that the dolphin can imitate aspects of the human voice. But the evidence Lilly puts forth does not support his implied deduction that the dolphin must have language. The explanation is likely much simpler.

Humans have good hearing, but in dolphins and other odontocetes the sense is absolutely exquisite. The ears detect an extreme range of frequencies, and they may well pick up aspects of sound that we cannot sense at all. The acoustic nerve, which bears sound information to the brain, has three to four times as many fibers as the same nerve in the human. The brain is even more elaborate. A human has about 25,000 brain neurons for hearing; the harbor porpoise, 650,000.

Unsurprisingly, odontocetes attend carefully to sounds. Karen Pryor, author of *Lads before the Wind* and a student of both dolphins and learning theory, says that dolphins relish complicated sounds the way dogs are enamored of complex smells. A dog finding a bit of smelly carrion will run its nose over every nub and corner of the offal, finding there a delicacy of odor that we with our dull noses will never know. Dolphins give equal interest to sounds. If the dog is the connoisseur of smell, the dolphin is the epicure of sound.

In addition, some of the odontocetes, like the bottlenosed dolphin, are talented vocalizers. Put an animal with sensitive ears and a talent for sound-making in an experiment where it is supposed to listen carefully and make sounds, and, what do you know, the animal listens and sounds off. Lilly's reprogramming paper shows this and nothing more: a sophisticated listener and vocalizer listening and vocalizing in a sophisticated way.

As for the mimicry, odontocetes are talented all-round mimics. They can watch an action, often only once, then repeat it with stunning accuracy

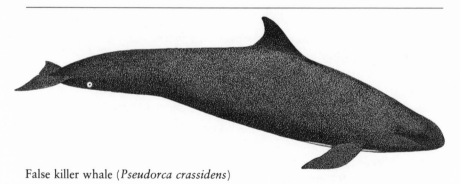

False killer whale (*Pseudorca crassidens*)

the first time. The spinner dolphins get their name from their full-twisting leaps into the air. The bottlenose never leaps this way in the wild, but captive bottlenoses have been known to watch a spinner spin and then perform the selfsame leap. A pregnant beluga whale at the Vancouver Aquarium was in the habit of resting on the bottom of the tank where wall met floor below a rank of viewing ports. There she lay, belly up, flippers out. The other two belugas in the tank, a juvenile female and an adult male, quickly took to resting on the bottom in the same belly-up, flippers-out posture, the head of one against the flukes of the other, like three great white peas in an aquatic pod. At the oceanarium in Port Elizabeth, South Africa, C. K. Taylor and G. S. Saayman played purposefully with the physical mimicry of the Indian Ocean bottlenosed dolphin. Dolphins put in the same enclosure with Cape fur seals imitated the sleeping posture and other movements of the seals, some of them sexual actions. One dolphin watched a diver clean an observation port, then repeated the diver's movements, all the while making a sound like air bubbles escaping the diver's regulator. Yet another dolphin watched a diver use a mechanical scraper to clean algae from the water flow into its tank, and, when given the scraper, used it just as the diver had.

And that is only the beginning of the dolphin's feats of imitation. Sea Life Park in Hawaii once had two female rough-toothed dolphins, Malia and Hou, who lived in holding tanks off the main tank, where they gave public shows. Each dolphin had been trained separately at quite different tricks. One day, unknown to the woman who ran the public show, the dolphins were inadvertently switched and put into the wrong holding tanks. So when the show routine called for Malia, Hou was released instead. She seemed upset, but she made it through Malia's tricks, not perfectly but well enough so that the audience suspected nothing. Next it was Malia's turn, and she, too, pulled off the other dolphin's tasks. From watching alone, the two dolphins had learned enough to improvise and imitate their way through the unfamiliar stunts.

Toothed whales often gave their mimicry a vocal twist. Karen Pryor surveyed dolphin and whale trainers and uncovered many instances of vocal mimicry, mostly in bottlenosed dolphins, orcas, and false killer whales. Two of the bottlenoses were devoted mimics that would parrot anything and played mimicry games with the single-minded devotion Pete Rose gives to baseball.

The mimicry no doubt has an underlying cause in the facts of marine life. Vocal mimicry exists also in birds and monkeys, creatures that live in groups in treetops where they cannot see one another and can easily lose contact with their fellows. These animals, like dolphins, are prodigious sound makers. If each bird or monkey imitates the sounds of the other birds or monkeys in its group, all the sounds of the group will converge into a common sort that allows the members of the group to identify and keep track of one another even when out of sight. The dolphin likewise lives in an environment where sight extends but a fraction of the distance over which the group may be spread. Common sounds reinforced by mimicry allow the group to stay in touch in murky and traceless reaches.

So, when you cut away all the puffery and the inflated words from Lilly's paper, the leftover is none too earthshaking — vocal mimicry by an accomplished vocal mimic.

The story Lilly tells about the orcas that bothered the Norwegian fishing fleet in the Antarctic bears a second look. Lilly argues that the orcas must have had language to escape the catcher boats, but the explanation need not be so ornate. To survive, animals have to be able to recognize danger and do what they need to stop it or escape it. That recognition requires neither language nor even an understanding of the danger beyond the fact that it is dangerous. It is a commonplace of deer hunting that the best chance for a buck comes the first hours of opening day, because once a shot is fired, the deer become very wary. The deer have no language; they pass no tales of firesticks through the woods; they surely know nothing of either the fish and game calendar or of ballistics. But they do know, in whatever way deer know, that danger is about. Whales are no less adept than deer, and the orcas avoiding the catcher boats were simply avoiding danger. They did not have to talk to do that.

Quite apart from Lilly's unwarranted deduction about language, the details of the story seem odd. The number of orcas far exceeds the largest reported aggregation. It also is very curious that the Norwegians, pioneers of the modern catcher boat, would have been using converted corvettes.

Nor does it seem likely that the whaling fleet, competing for its share of the Antarctic whaling quota, would have sacrificed catcher boats just to chase off some aggravating orcas.

An inquiry to the Royal Ministry of Fisheries in Norway brought an interesting reply. Since World War II, Norway has not fished the Antarctic, with the single exception of the 1951–1952 season, when a "fleet" comprising two or three ships operated experimentally near South Georgia. There were no Norwegian fishing vessels in the Antarctic in the year Lilly gives for his story, and there hadn't been for several years. According to Ernst Aas, director of the ministry, no one in that agency has heard of the encounter Lilly reports — striking ignorance of so dramatic an occurrence.

Then there is the matter of what Lilly calls the cetacean ethic, which dictates that humans are a special species and that cetaceans shall not hurt them. In truth, for predatory animals of their size, odontocetes are extremely gentle. They could inflict serious injury, even kill if they wished, but they do not. Nor do cetaceans have to be tamed before they can be handled safely. Gil Hewlett, curator of the Vancouver Aquarium, has studied orcas in the wild and kept them in captivity. He calls them the "most efficient predator on the earth." But, he says, "when you get them in a net, they're just like big pussycats. You can get right in with them and they won't hurt you."

Impressive and intelligent restraint from an impressive and intelligent predator, but by no means evidence of ethics. The odontocetes that do well in captivity, like the orca and the bottlenosed dolphin, are social predators. They can be violent but they live in groups that violence can destroy. Restraining violence is a crucial part of social life, in odontocetes as in humans or wolves, and that restraint is a crucial part of the inheritance of all such social predators. To survive in and sustain dolphin society, a dolphin must pull its punches. With a human, it is acting toward the human as it would to another dolphin, not holding to some cetacean ethic of special treatment.

But Lilly goes further. He asserts that no captive cetacean has ever harmed a human. That simply isn't true.

According to Karen Pryor, adult bottlenoses typically assert dominance over humans from time to time, much as timber wolves raised with people periodically take the upper hand with main force. A dolphin bent on dominance will butt and ram without apparent provocation and may invent some aggressive teasing game, like roughing you up and then fluke-slapping you out of the water. Sea Life Park had one female bottlenosed dolphin that never stopped doing this sort of thing and finally became downright dangerous, ramming or slapping anyone who got in the water with her. To avoid adverse publicity, oceanariums have been reluctant to admit problems with large whales, particularly orcas, but there have been incidents. At Sea Life Park, a diver made the mistake of taking an empty fish bucket from a false killer whale that was playing with it. The whale seized him in its jaws and pinned him to the bottom for some minutes, opening a cut in his ear that required stitches. Pryor also knows an unnamed source who saw a graver incident at Marine World near San Francisco. During rehearsal of a head-in-the-mouth routine, an orca took the trainer to the bottom and worked him over.

Given the number of cetaceans in captivity and the artificial, even brutal, conditions of their confinement, it is actually remarkable that so few attacks have occurred. Indeed, the cetaceans' restraint and patience toward humans is remarkable. But Lilly's claim that no captive cetacean has ever hurt a human is groundless.

Lilly's orca story and the unsubstantiated claim about the absence of violent incidents with captive cetaceans fail to stand up to ordinary journalistic scrutiny. It is as if the ideas already were there and the facts came along later to support them. Were they isolated examples, they would mean little. But the pattern is repeated, and they mean much.

Like anyone interested in whales, I grabbed at the chance to hear Lilly speak, in this case at the College of Marin, north of San Francisco, in November 1977. Lilly wore black, looked sallow, gave a technical lecture on brain damage by rotatory acceleration and frequency changes. His voice was as flat as his material. The woman next to me nodded off. The interesting part came after the lecture, with the questions and answers.

Asked whether orcas ever prey on other cetaceans, Lilly said the only case he knew of was a captive killer that set upon an old bottlenose. He intimated that this was euthanasia. Then another questioner said that a film made in Puget Sound showed orcas toying with a newborn harbor porpoise and then apparently eating it. Lilly smiled indulgently. His face had a don't-bother-me-little-girl look to it. Orcas, like humans, have their "bad boys," he said.

The evidence that at least some orcas attack other cetaceans is ample and readily available. For example, a photograph taken by Camille Goebel of two orcas buffeting a newborn harbor porpoise, probably the same incident Lilly's questioner described, appeared in the May 1977 issue of *Pacific Search*. Why Lilly chooses to overlook such evidence and then, when confronted with it, to explain it away in terms of moral degeneracy only he knows.

In response to another question about orcas, Lilly claimed that the number of attacks by killer whales on yachts and small boats has increased in recent years. The significant point about these attacks, Lilly continued, was that once the boat was sinking and the people safely off into the life rafts, the orcas ceased the assault. The orcas were trying to tell us something, and they must leave survivors to bear the message.

The truth of the matter is that there has yet to be a single verified attack by a killer whale on a small boat. The supposed attacks that occur now and again are invariably collisions with sleeping whales. Killer whales do not attack yachts to begin with, much less have they taken to assaulting boats more often.

Then a man up front asked the inevitable question about cetacean intelligence: If the whales are so smart, why haven't they learned to avoid whaling ships? The answer, unfortunately, is pedestrian. To get the quantities of food needed to support their vast bodies, whales must feed in certain highly productive ocean areas at those times of the year when production is the highest. The whalers simply hunt the feeding grounds during the appropriate season. The whales have no alternative source of food. They either stay away and starve, or they feed and risk death by harpoon. The question has nothing to do with intelligence; it has everything to do with necessity.

This stark and true answer, though, was not the one Lilly gave. He hitched up his belt; he stared down at the questioner. "Apparently," he said, "you don't have all the facts." Once whales were common all around, Lilly said. They even came up the Hudson River. Then, when whaling started, the whales disappeared. By 1845, the whales had vanished, or so the whalemen thought. What the whalemen didn't know was that the whales had retreated to the Antarctic and were hiding out. The whalemen didn't find them until 1925, and then they invented the steam catcher to go into the Antarctic and kill the whales. Thus, Lilly concluded, the whales have tried to escape us and our destruction, but they have no alternative left save flight, of which they are physically incapable.

The crux of Lilly's statement is that the whales retreated south from the whalers. The Yankees killed sperms, rights, and bowheads. The Antarctic whalers killed rorquals, which the Yankees never pursued. The Antarctic whalers rarely killed sperms or rights, because there were so few of them. In short, the whales the Yankees pursued weren't the same whales the Antarctic whalers went after. Lilly's scenario of a strategic global withdrawal translates into the marvelous assertion that the sperms, bowheads, and rights transformed themselves into rorquals. Biology has one law you can always count on: kittens don't grow up to be rhinoceroses, and bowheads never become blues. Lilly's implication that they did is nonsense.

In answer to yet another question, one about the relation of his isolation research to the work on dolphins, John Lilly gave an answer that may be the hint of an explanation. Lilly said that he did his first experiments with isolation in 1954, the first cetacean study in 1955. "The one was a consequence of the other," he said. "The two have gone in parallel." He might better have said that he is unsure where the one begins and the other leaves off.

Lilly spent several years exploring the outer reaches of human consciousness. He was convinced that there lay salvation, an end to the evil and pain of the human condition. He writes in *The Center of the Cyclone*, a chronicle of his spiritual journey: "It is my firm belief that the experience of higher states of consciousness is necessary for the survival of the human species. If we can experience at least the lower levels of Satori, there is hope that we won't blow up the planet or otherwise eliminate life as we know it. If every person on the planet, especially those in power in the establishments, can eventually reach higher levels or states regularly, the planet will be run with relatively simple efficiency and joy. Problems such as pollution, misuse of natural resources, overpopulation, famine, disease, and war will be solved by the rational application of realizable means" (p. 3).

Dolphins enjoy such an improved state, and humans can profit mightily by living according to the dolphin way. Lilly told the participants in his workshops at Esalen: "The dolphins were close . . . they had a freely moving and joyful life constantly together and . . . they had no compunctions about bowel movements, urination, or their sex life. I intimated that humans could well afford to try this way of living. . . . Today I still feel that by emulation of the dolphins' ways we could make a lot faster progress in loving one another and in enjoying life and abolishing tensions that go on between groups of people" (*Center of the Cyclone*, pp. 105–106).

But, remember, the dolphin is a mental midget next to the sperm whale, which has the biggest brain on the planet. The sperm must use little of that great mass of gray matter on such mundane day-to-day matters as food and sex. The remainder of its giant brain deals with inner experiences: "If a sperm whale wants to see-hear-feel any past experience, his huge computer can reprogram it and run it off again. His huge computer gives him a reliving, as if with a three-dimensional sound-color-taste-emotion-re-experiencing motion picture." Humans can accomplish similar operations only with artificial computers. The sperm whale, though, is no mere mechanical self-programmer. "The sperm whale probably has 'religious' ambitions and successes quite beyond anything that we know. His 'transcendental religious' experiences must be quite beyond what we can experience by any known methods at the present time. Apparently, we can in rare times with our experience begin to approach his everyday, accomplished abilities in the cognitive, conative, emotional spheres" (*Lilly on Dolphins*, pp. 218–219).

This picture of the dolphin as a graduate of Esalen and of the sperm whale as the greatest transcendental meditator of all is surely Lilly's greatest failure. John Lilly was teaching at Esalen and beginning the climb toward higher states of consciousness. So, Lilly writes, the dolphin has already realized the Esalen life-style and the sperm whale commutes daily to complete transcendence. This isn't even speculation. This is autobiography masquerading as science.

According to the schema by which academicians divide the known world, intelligence belongs to the psychologists. They, unfortunately, have not done well by the concept. So far, no psychologist has come up with an acceptable, essential definition of intelligence. That is, the psychologists don't know what intelligence really is. But they do, or so they say, know how to measure it with IQ tests. The psychologists have settled for an operational definition: intelligence is whatever IQ tests measure.

The trouble with this definition, quite apart from its circular logic, is that it tells us little about the nature of our own intelligence and nothing about the intelligence of other species. We have to rely on common sense.

Our commonsense understanding of intelligence has something very much to do with communication. When we say that a particular animal is intelligent, we mean much the same as "This animal communicates in a sophisticated and varied way." A female moth ready to breed releases a chemical signal, called a pheromone, that excites and attracts male moths downwind. This is communication, make no mistake, and it is effective. It is also simple and unvarying, not a way of conveying individual experience. Dogs send other dogs and observant humans considerably more complex messages that often have an individual mark. Language, though, is the most sophisticated way of communicating group and individual experience. Language was central to the evolution of the human species and the human brain. We are group animals, group predators; language was the principal weapon in the hunt and, perhaps even more important, the symbolic glue that kept the band together. Reasoning from our own abilities and history, we equate sophisticated communication with intelligence.

John Lilly's ideas about cetacean intelligence follow the same logic. Brains of the size of the cetaceans' must mean intelligence and, therefore, language. Unfortunately, Lilly's sole demonstration of language is the vocal reprogramming paper, which delivers much less than it advertises. But that need hardly be the last word on the topic.

Yet it very nearly is, largely by default. John Lilly's ideas, however poorly presented and supported, are unsettling. You might think that those ideas would prompt other scientists to study dolphins. But the past ten years of *Psychological Abstracts,* which summarizes research from American and foreign scholarly journals, yield fewer than a dozen studies of dolphins, compared with thousands on rats. Apart from navy employees studying marine mammals, the only psychologist with cetaceans as his principal research interest is Louis M. Herman of the University of Hawaii. Cetaceans could prove to be where the real nonhuman psychological action is, but for the moment the psychologists are playing it safe with their rats, running them through mazes and Skinner boxes and teaching them various ratty feats. Of such trivial accomplishments are academic careers made.

Because cetaceans are so little used as experimental animals, there is little experimental information that bears on the question of intelligence in dolphins and whales. Two experiments deal with the ability of cetaceans to reason. L. V. Krushinskii, B. A. Dashevskii, N. L. Krushinskaya, and I. L. Dmitrieva of the Institute of Developmental Biology of Moscow State University ran tests to determine how well two female bottlenosed dolphins could coordinate spatial figures without learning. The investigators felt that their results showed elementary reasoning in the dolphins. In the second study, Louis Herman and William Arbeit tested the learning efficiency of bottlenosed dolphins in hearing-discrimination problems and got results comparable to those for advanced monkeys and apes. These studies, slim as they are, show that the dolphin has a good working brain, perhaps good enough to communicate in sophisticated ways. But does the dolphin communicate in such ways? Three experiments, all of them conducted under the auspices of the navy's marine mammals program, address that central question.

In the first experiment, designed by Thomas G. Lang and H. A. P. Smith, two bottlenosed dolphins named Dash and Doris occupied separate tanks connected by an intercom. During the experiment the intercom was turned on, then off for periods of random length. When the intercom was on, the dolphins whistled much more than when it was off, and they whistled back and forth, with one animal remaining silent and the other vocalizing. But Lang and Smith could find no pattern to the whistles. As far as they could tell, Doris and Dash were simply saying hello back and forth over and over again.

The second experiment was set up by Dwight Wayne Batteau, an engineer of wide knowledge and talent, who made a device to translate human vocal sounds into dolphinlike whistles. To get distinct and unambiguous whistle contours, Batteau used artificial words, some of them Hawaiian. PLOP meant "hit flukes on water"; MAUKA, "come to station"; RE-PEAT, "whistle word just given." Two bottlenosed dolphins served as subjects: Dash, rechristened Maui for the sake of whistle contour, and Puka. Batteau gave Dash or Puka a command beginning with the dolphin's name, rewarded a correct response with a fish and a wrong one with NEG-ATIVE — whistled, of course. Dash showed more aptitude than Puka; he responded accurately to about fifteen words.

But responding to words is by no means the same thing as understanding them. You can train a dog to come to a snap of your fingers, a toot on a whistle, two taps of your right foot, or the word *come*. The signals are arbitrary; *come* conveys no more meaning than the finger snap or the toe tap. Batteau's experiment produced no evidence that to the dolphins the whistled words were different from finger snaps. For example, Batteau couldn't give Dash or Puka two commands at once. If he told one of them to hit the ball with a flipper and swim through the hoop, the dolphin only hit the ball, as if there had been no command about the hoop. If the experimenter changed props and gave a new command, the dolphin responded incorrectly or with the action associated with the object. Dash had learned to swim through the hoop. If he was told instead to hit it with his flipper, as he would the ball, he either did something wrong or swam through the hoop instead of hitting it as the command told him. The dolphins obviously learned to perform, but they never appeared to perceive words as words.

The third and most interesting experiment was designed by Jarvis Bastian, a psychologist and psycholinguist from the University of California at Davis. The purpose of Bastian's experiment was to see whether one dolphin could send information to another. The dolphins were put into a large tank and separated by a curtain, so that they could hear but not see each other. Buzz's side of the tank had two paddles. For both dolphins to earn a food reward from the automatic fish feeder, Buzz had to push the correct paddle. But only Doris knew, from either a flashing or a steady light on her side of the tank, which paddle was correct. For her to get a fish, she had to tell Buzz which paddle to push.

Bastian started the experiment certain that dolphins lack language, and it surprised him to no end that in little time Buzz was picking the correct paddle nine times out of ten. However, it took the experimenters a long time to connect Doris's sounds to Buzz's choice. Finally they found that Doris was making a short series of clicks when the light held steady and keeping quiet when it flashed. This raised a new question: were Doris's clicks voluntary or were they randomly associated with the lights?

Bastian put together another set of experiments and found that Doris clicked even when she knew Buzz was missing from the tank. As Bastian explained it, Doris had learned unconsciously and quite by chance to make certain sounds in response to the light and Buzz had learned by trial and error what the sounds meant. Behaviorists call this adventitious conditioning. It was not communication.

Since these three experiments, research had moved away from uncovering delphinese to finding ways for humans and dolphins to communicate. For example, William Langbauer and his associates at Florida's Marineland are teaching dolphins to manipulate symbols to create a language, much as has already been done with chimpanzees. Louis Herman of the University of Hawaii taught a twelve-word sonic vocabulary to a dolphin, who was showing some understanding of simple sentence structure and was nearly ready for two-way conversation. Unfortunately, the research ended when the dolphin was taken from the laboratory without Herman's knowledge and released into the sea. Lilly's Project JANUS is of the same basic ilk, entailing a code of 64 sounds transformed by computer into the hearing ranges of dolphin and human.

Still, the final experimental word has hardly been given on whether dolphins actually communicate complex information to one another. None of the navy's experiments was definitive.

The Lang and Smith study rests on the basic assumption that the intercom faithfully rendered the dolphins' signals. This is a big assumption. We know that dolphins have exquisitely sensitive hearing and that they pick up aspects of sound we are barely if at all aware of, yet we have little idea which aspects of sound dolphins respond to. An intercom that to human ears renders the highest of high fidelities might to a dolphin sound like a scratchy, coughing long-distance telephone connection. The whistles exchanged between Doris and Dash may have meant something like, "Hello, Dash, is that you?" "I can't hear you." "What was that? This line is lousy." "Would you please speak up?" and so on.

Batteau's human-dolphin translator involves the same assumption. The whistles the translator gave off sounded dolphinlike to the human ear, but there is no necessary reason to believe that they sounded like a dolphin to another dolphin. As for the dolphins' failure to respond to words as words, the design of the experiment assumes that words to a dolphin are the same as words to us. Unless the translated whistles happened to copy the same sound characteristics and arrangements dolphins use to make words, they would respond to the translator's whistles as cues, not as words — reacting to them as if they were Bastian's lights, not vocal symbols bearing distinct meaning. We humans could, if we stray into the wrong intergalactic neighborhood, find ourselves in an analogous fix. To us a bell is merely a signal. It rings, and we stop teaching or go to lunch or say the Angelus or do whatever a bell means in that situation. It is simply a signal, like snapped fingers to the trained dog. Imagine what would happen if we were to come upon aliens whose "language" comprised frequency- and amplitude-modulated bell ringing in a range partly exceeding our hearing limit. When the aliens tested us for bell-ringing language, they would discover, lo and behold, that we are bell-speechless. The dolphins might be in the very same predicament.

Bastian's experiment avoids this pitfall by letting the dolphins use their own sounds, but his procedure also falls short, simply because the dolphins proved more complex than they were thought to be. The experiment simply didn't hold the world as steady as it needed to. As Buzz and Doris showed by solving their problem with chance learning, important variables remained outside Bastian's control. The study is too full of ghosts, holes, and hindsight to support any statement as pontifical as "Dolphins do not communicate intelligently."

The problems with Bastian's experiment point up a general issue: Just how much should we trust any experimental investigation into the minds and brains of animals? The laboratory situation is artificial and, ultimately, arbitrary. It suits the need of keeping the animals, who are often unwilling subjects, securely captive under unnatural conditions, and it serves to fit the form of the results to the epistemological dictates of science. We have to assume — or hope — that the experiment replicates the world of its animal subject closely enough that whatever happens means what we claim it does. This calls for an act of faith every bit as grand as belief in virgin birth or the rediscovery of the ten lost tribes in Utah. Since we live in a culture that pays obeisance to science, we invest such faith more easily than we should. Our trust nearly violates common sense. There is no reason to believe that an apparently bright and inventive animal, whose natural habitat is the open sea, can be studied accurately in a laboratory by a Ph.D. whose life revolves about the university library, a research facility, and an academic department. In testing the "intelligence" of other animals, we are something like the early IQ testers who gave their exams in English to non-English-speaking immigrants in the disturbing environment of Ellis Island and, from the immigrants' poor results, pronounced Eastern Europeans undesirably dull. The dullards were the testers themselves, who were too thick-headed to see their "science" for the pomposity it really was.

Experiments are never enough, particularly in an instance such as this, where they are so few. We must understand the animal as it is in its own world. We have to see what makes a dolphin a dolphin or a whale a whale and then decide whether "dolphinness" or "whaleness" shares anything with what we call intelligence.

Cetacean social structure, like the span in size from the harbor porpoise to the blue whale, covers the gamut from anarchic and solitary to organized and highly social. A few species, like the pygmy killer whale, live out their lives all alone. Others have loose social groupings that change during the course of a year. Rorquals, for example, live in small pods, as whale groups are called, of from two or three up to about twenty animals, and these small pods coalesce into much larger groups of hundreds and even thousands of animals on the seasonally rich feeding grounds.

We could be overlooking part of the social reality of these large whales by merely looking. Usually, "herd" or "group" means a number of animals clumped together close enough to be within eyeshot of a human observer. Sound, not sight, may connect whale herds, and animals seemingly scattered over many square miles of water may in fact be traveling or feeding together. Roger Payne and Douglas Webb have made a fascinating speculation about this possibility. Finback whales emit pure 20 Hz sounds, about an octave deeper than the lowest key on the piano, that are almost as loud as the roar of an airliner lifting off the runway at close range. In water such deep sounds carry a long way before dissipating, and the structure of the ocean itself, which is like a sheet and spreads sounds with reflections off surface and floor like a cylinder, carries them even farther. A 20 Hz sound made near the surface in a temperate ocean should be audible at 30- to 35-mile intervals over a distance of 400 miles and maybe more. In fact, since sound transmission in water varies with pressure and temperature and thus with depth, there are theoretical conditions under which the finback's 20 Hz signal could carry 11,500 miles in a straight line and spread through millions of square miles, a range greater than the expanse of any ocean. All the finbacks in one ocean could be in some sort of contact. It is unlikely, though, that the 20 Hz sound is anything more than a signal to indicate presence. The sound is too simple to convey complex meaning, and temperature and pressure differences over long distances distort the sound so much that any detailed information would be badly garbled. But the whales could use this sound to guide one another to a rendezvous site. Such a guiding mechanism could explain how finbacks, which breed in winter when food is scarce and the whales are widely scattered, ever find mates.

Toothed whale societies are easier to pick out because the cetaceans stay closer together. Sometimes these animals gather in vast congregations. A. G. Tomilin, a Soviet cetologist, reported a school of common dolphins in the Black Sea so immense that he estimated it at 100,000 animals. This was probably a seasonal grouping based on a sudden abundance of prey, much like the summertime aggregations of rorquals. Other dolphins gather in schools numbering from a few animals to several thousand. Little is known about the internal makeup of these groups because of the difficulty of studying them. In one of the few such studies, William Evans and Jarvis Bastian watched bridled dolphin schools and found three basic groupings: a lone male, sometimes accompanied by a single female; four to eight immature males; and five to nine adult females and young. This organization suggests a herd of large grazing animals, which comprises subunits divided by age and sex. The similarity is actually unsurprising. Ecologically, a big school of dolphins traveling through the sea is much like a herd of wildebeest making their way across the Serengeti.

In some of the odontocetes, social organization is elaborate and persistent. Twenty to fifty sperm whale cows and their young stay together in a tight nursery school presided over by one or more adult bulls, who are called schoolmasters and are nearly twice the size of the females. Younger bulls, too old to remain in the nursery and too young to have schools of their own, gather in bachelor pods. In migration, pods and schools may form up into bodies, as they are called, that sometimes number 1,000 whales. Sperm whale society is polygynous; the schoolmasters keep the females for themselves. As in other polygynous marine mammals, like the fur and elephant seals, the bulls brawl for position, fencing and biting with their jaws, and ramming into each other's bodies with their great bluff heads. The fights are bloody and violent. Moby Dick's twisted jaw was the leftover of such a struggle, and jaw injuries and broken ribs are not uncommon in fully grown sperm bulls.

The orca has long been regarded as the wolf of the sea, and the analogy is, by accident, most apt. Neither orca nor wolf is the bloodthirsty villain of legend, but both animals are highly and complexly social. The orca pod is an extended family of five to twenty animals. Usually, every five whales include one bull, one calf or juvenile, and three cows. As in sperm whales, the male orca is bigger than the female. The pod is cohesive; its members travel close together, often hunting and even breathing as a unit. A recent cooperative study by American and Canadian researchers identified four pods of about 70 orcas resident in Puget Sound, Georgia Strait, and Juan de Fuca Strait. These pods sometimes joined to form a superpod when the salmon were running, and nine pods of about 50 whales periodically entered from the sea and returned. The pods last. One pod group led by a distinctly marked bull regularly frequented the San Benito Islands off the coast of Baja California for some twenty years.

The highly social cetaceans are so social that alone they cannot or will not survive. In a study that almost went tragically awry, the crew of Jacques Cousteau's *Calypso* caught a female common dolphin and kept her in a net enclosure alongside the ship. For hours and hours the dolphin whistled the same dolorous call, then she fell silent, slipped into shock, and would certainly have died had not the men brought her aboard and hand-fed her. Returned to the enclosure, she was presented with a male companion. The silent female suddenly chattered like a long-silenced gossip, twisted and turned and leapt in paroxysmic joy, and led her newfound friend on a guided tour of every nook, cranny, and fold of their net-home.

Captive dolphins and other toothed cetaceans form such strong pair bonds that the loss of the beloved has been known to kill the lover with ulcers. Cetaceans likewise protect and aid members of their group. A predator attacking a cetacean may find itself set upon by the cetacean's companions. Sperm whale schoolmasters often defend their harems vigorously, and not a few whalemen in the era of wooden whaleboats went to the bottom because they had struck the cow of a particularly stalwart bull. Even these days, with all the odds on the side of the steel-hulled catcher boat and its harpoon cannon, the bulls occasionally counterattack, succeeding only in making themselves the gunner's next target. Usually, though, cetaceans render assistance rather than return the attack. In an unfortunate accident, the *Calypso* overran a sperm whale calf, which hit the propeller and was hurt mortally. The calf's cries brought 27 sperm whale cows, who attended the calf until it died. Dolphins will support a hurt companion, bringing the wounded animal up to the surface to breathe. Farley Mowat, in his frightening *A Whale for the Killing*, says that the finback male who was the consort of the female trapped inside a tidal lake stayed close for weeks, often breathing in synchrony with his mate, leaving only when she died.

The strong pair bonding and fraternal care in the cetaceans is, however, not unusual, nor are the kinds of social organization seen in whales and dolphins markedly different from the societies of other mammals. The same sorts of social units occur in other families and species, and the care and concern cetaceans show for their fellows, while it strikes altruistic chords in our own hearts, have their equal in other social mammals.

Most of the known behavior of cetaceans fits into our present under-

standing of how animals act. But there are a few oddities and individuali-
ties that should spark far more curiosity among certified scientists than
they appear to.

Cetaceans cooperate with humans willingly. Pliny the Elder (A.D. 23–79)
wrote that dolphins worked with humans to drive mullet into nets in the
mouths of the Rhône River near Nîmes, France, and that dolphins herded
fish toward fishermen on Iassos who fished with spears by torchlight at
night. Oppian, a later Greek poet, also reported torch-fishing at Euboea.
Until only a couple of decades ago, scientists derided such stories as pure
nonsense. Admittedly, neither Pliny nor Oppian qualifies as an unimpeach-
able source, but the scientists' own beliefs kept them from even entertain-
ing the notion that such reports might be true.

Then the scientists found out that such cooperation between humans
and cetaceans occurs yet. On the coast of Mauritania, fishermen slap a
rolled cloth on the water to call the dolphins to herd running mullet into
their nets. Aborigines at Point Amity in Queensland, Australia, beckon
helpful dolphins similarly, by striking the water with the flat of a spear,
and fishermen and dolphins work together. F. Bruce Lamb met a Brazilian
named Rymundo Mucuim who spearfished on the Tapajós River with a
boutu, or Amazon river dolphin. Mucuim summoned the dolphin by whistling and tapping a paddle against the side of his canoe. The *boutu* pushed
fish into the shallows where the man could see and strike them, and it
caught for itself the ones that escaped him.

Anthropologists sometimes theorize that the long-standing arrangement
between humans and dogs began when a pack of wolves and a band of
humans found they could both eat better by hunting together. To my
knowledge, though, no such cooperation occurs today between wild
wolves and human hunters. Yet dolphins and fishermen continue to work
side by side.

Unique though this behavior is, it can be explained as learning based on
a payoff. The humans help the dolphins, just as the dolphins help the
humans. But sometimes cetaceans cooperate or associate with humans
even when there appears to be nothing in it for them. Scientists also used
to deride the stories about dolphins carrying shipwrecked and drowning
people to safety ashore. Some of these stories are the hallucinations of
people who have been pulled back from the threshold of death, but others
have the ring of truth or the corroboration of eyewitnesses. Dolphins, do,
of course, support their own kind in distress, and in rescuing a troubled
human they are simply extending that behavior to another species. Yet
how does the dolphin recognize that a human, a species it rarely encoun-

ters, is in trouble, and how does it figure out that taking this unknown creature ashore is the right thing to do?

Many Greek and Roman writers, such as both Plinys, Oppian, Phylarchus, Plutarch, Athenaeus, and Aelian, told stories, some original and some borrowed, of dolphins that willingly struck up friendships with humans. As with the stories of cooperation and rescue the scientists poohpoohed such tales until a dolphin proved them mistaken. In 1955 a female bottlenose regularly visited the beach at the New Zealand resort town of Opononi. Dubbed Opo, she at first preferred children, later learned to accept adults, played elaborate games of ball and tag, and even romped in the shallows with a dog whose reactions, could they be known, must have been priceless. Opo attracted hordes of tourists and became a national figure practically overnight. Unfortunately, like many animals that venture too close to civilization without knowing its hidden dangers, Opo was killed by mishap.

It remains to be explained why Opo sought people out in the first place. She was alone, so perhaps she was seeking companions to replace the pod she had lost or left. But why didn't she try to find other dolphins? Plutarch wrote that the dolphin, alone of all the animals including humans, seeks friendship at no advantage to itself. Perhaps, in delphinid terms, friendship is its own reward. That is certainly true to some extent of humans, chimpanzees, and dogs. Perhaps, too, the dolphins simply find us interesting.

When Forrest Wood was curator at Marine Studios, a female bottlenosed dolphin was captured and added to the colony. Named Priscilla, she adapted very fast to life in the tank, taking food from the hand almost immediately. Priscilla had a friendly and winning personality, but she proved a poor learner and she was included in the six animals given to Lilly and the brain scientists for experiment. By luck of the draw and Priscilla's charm, the scientists decided against sacrificing her, and she was spared martyrdom on the altar of neuroanatomy. Wood decided that the oceanarium couldn't afford to keep feeding a dull dolphin, and he had Priscilla returned to the wild. Eighteen months later, after an epidemic of erysipelas ravaged Marine Studios' dolphin colony, a capture team hunting replacements came in with two female bottlenoses, which, surprisingly, had run aground and let themselves be captured without a struggle. When put into the tank, the larger of the two acted like someone who had been there before, and the handlers suddenly realized that Priscilla was back.

For a dolphin dismissed as dull, Priscilla proved inventive enough to outwit the very folks who had dismissed her and to get where she wanted to be. Captive odontocetes put that same inventiveness to their play. They play most when young, less as they age, but even as adults they seem to enjoy play for the sake of play. They make up their own involved games like hide-the-fishbucket or chase-the-feather. They tease mischievously. They have been known to harass groupers and moray eels for the sheer devilry of it, to pick handkerchiefs from unguarded pockets, and to take such an active dislike to Roman Catholic priests to spit stones at anyone wearing a clerical collar.

Karen Pryor and Ingrid Kang decided that their show at Sea Life Park had gotten altogether too slick and, to liven things up, they hit upon the idea of showing the audience how training works by teaching Malia a new trick at each show. Malia was a good subject; she was a rough-toothed dolphin, a species with extreme curiosity, short temper, long attention span, and a love for puzzles. Malia learned new trick after new trick until Kang and Pryor were having trouble coming up with novel routines. Malia was undaunted. She started doing stunts, like flinging herself from the water backward and leaping upside down, that she had not been taught and had never done before. Working at Sea Life Park at the time was Gregory Bateson, Renaissance man and all-purpose genius, who felt that Malia had put facts together to abstract the principle that novelty would earn a reward. This pointed toward sophisticated learning. At Bateson's excited urging, Pryor started over with a new subject and recorded everything to determine whether and when the dolphin, another rough-tooth called Hou, caught on to the notion that the desired behavior was new behavior. Unfortunately, Hou was a placid animal who, because frustration never got to her, did nothing new that the trainer could reward. Then, over several sessions, Hou saw the light. She shed her placid ways and became an excited subject, pulling off new stunt after new stunt. Pryor and her assistants, Richard Haag and Joseph O'Reilly, published a scientific paper titled "The Creative Porpoise." The results certainly justified the title. In the introduction to Pryor's *Lads before the Wind,* Konrad Lorenz writes: "To grasp the fact that it is not any particular pattern of behavior which is going to be rewarded, but that the desired action should be some form, *any* form of movement which has *not* as yet been reinforced, requires a feat of abstraction quite unexpected in any animal."

Yet this experiment shows just how difficult it can be to interpret the meaning of an animal's actions in captivity and in experiment. The creative-porpoise experiment would seem to show that dolphins must be highly intelligent. But Pryor duplicated the novelty effect with distinctly duller subjects: pigeons. Put through the same kind of training as Hou, the pigeons came up with such unpigeonish displays as lying on the back, standing with both feet on one wing, or hovering two inches off the ground.

We understand little of the playfulness, ingenious or otherwise, of the mysticetes. On the whole, they appear duller and more stereotyped than the odontocetes. Mysticetes are grazing animals. They do eat animal life, namely zooplankton, but that life appears in huge plantlike blooms to be grazed, not hunted. The life cycles of most mysticetes revolve about the sea-graze: migrating toward the pole in summer to feast on the summer burst of life, returning to temperate or tropical waters in winter to breed and calve and scrounge for what food there is. Ecologically, a blue or finback is something like a deer or a buffalo. But the odontocetes are predators; they run down and kill individual animals. Because they must often outsmart their prey, predators are by and large brighter than the browsers or grazers they prey upon. This generality appears to hold as true of cetaceans as it does for other classes of mammals.

Yet, for browsers, some mysticetes are remarkably playful, particularly as calves and juveniles, and some of that capacity for games lingers into adulthood. Humpbacks and grays, in particular, cavort and leap for no apparent purpose other than cavorting and leaping, and they sometimes allow humans to approach, apparently to satisfy their own curiosity or to play with the boat that brings the visitors. In fact, mysticetes may play games that we do not recognize as diversions, particularly ones involving sound. William Schevill taped a humpback making echoes in a submarine canyon in the Bahamas. The whale made a sound and waited for the echo. Then it made the same sound one note higher, again waiting for the echo. Up and up the scale the whale worked, until it reached its own top note. Then it tried new sounds, making various gurgles and roars, always waiting for the echo before trying the next noise. The humpback kept up this elaborate diversion until a group of distant whales called and it swam off to join them.

And there is the final, great anomaly of the song of the humpback. These songs lack the melody and rhythm of music, but they have an eerie, haunting beauty to them. They are so memorable and suggestive that recordings of humpback songs were included in the payload of the Voyager satellite that is even now traveling away from our solar system out into the galaxy. The function of the songs, despite long research by Roger and Katy Payne, remains uncertain. The whales sing only in winter, and only the males participate. The females apparently are silent. All the males sing the same song, but the song changes as individual whales make up new phrases, which other whales copy. When the whales leave for the summer feeding ground, they stop singing, but when they return the following winter, they pick up right where they left off, as if the intermission had lasted fifteen minutes and not six months. Since only the males sing, it is likely that the song plays some role in the breeding ritual, as many bird songs do. But bird songs persist; they do not change from season to season as does the humpback's plaint. And why is the humpback song so complicated and so long? No one knows the answer. The mystery remains.

The question of cetacean intelligence comes full circle. John Lilly began with the cetacean brain. There now we must return.

The evolution of the vertebrate brain began in the simple chordate animals, like the fishlike lancelet *Amphioxus*. The nerves, which in the insects and mollusks encircle the gut, instead coalesced in the neural tube, which in time the spine enclosed. The head end of the neural tube protruded from the bone of the spine, and, lying close to the developing sense organs, it grew to act as an information processor. The growth took the form of three distinct bulges, called the forebrain, the midbrain, and the hindbrain.

These swellings developed into other, complex structures. From the hindbrain, for example, came the cerebellum, which regulates muscular coordination and balance. From the forebrain arose the cerebrum, a two-hemisphered structure that eventually became the centerpiece in the evolution of the vertebrate brain — at least from where we humans stand. In fish and amphibians, the cerebral hemispheres are small, devoted largely to smell. But in reptiles the cerebrum is more prominent, and in certain advanced species the forward upper parts of each cerebral hemisphere are covered with a thin skinlike patch of nerve cell-bodies called the neocortex. As mammals evolved from reptiles, the neocortex — the so-called gray matter — grew backward over the white matter of the cerebral hemispheres. In rats and shrews the neocortex is small and smooth. As the mammals evolved further, the neocortex added size and wrinkles, which increased further the surface area to make room for greater and greater numbers of nerve cells. In dogs, monkeys, and humans, the neocortex covers a very large portion of the entire brain, and its convolutions are the brain's most distinguishing feature. Think of "brains," and you think of that wavy surface.

Specific patches of neocortex receive incoming sensory messages, and others send out motor instructions to the body. In mammals with small neocortices, the motor and sensory patches take up almost the whole brain surface, but these same areas in the brain of a monkey occupy only about half the area, and in a human the proportion is smaller yet. In fact, accounting for differences in overall size, the sensory and motor areas in humans are about the same size as those in a marmoset or a gibbon, even though the cortex as a whole is much larger. This "extra" area is called the association cortex because it has the task of storing, sorting, associating all the many varieties of sensory and motor information. The association cortex gives us speech, reason, a sense of history, the capacity to imagine the future. At the risk of being aphoristic and simpleminded, one could say that the neocortex makes mammals mammalian and the association cortex makes humans human.

The cetaceans are the most divergent and specialized mammalian order, and their brains, although following the general mammalian plan, are unusual. There are three striking features about the organ. For one thing, the typical mammalian brain is longer than it is wide, but the brains of whales and dolphins are greatly foreshortened and thus wider than they are long and quite high. The foreshortening is more marked in odontocetes than in mysticetes. Second, the cerebellum is massive, making up about 20 percent of the total weight of the brain in the mysticetes and a shade less in the odontocetes. The cerebellum is over half again as large in cetaceans as it is in terrestrial mammals, where on the average it accounts for about 12 percent of the brain's weight. Cetaceans are remarkably graceful and well-coordinated, and the source of that striking grace lies in the large cerebellum.

The third — and most striking — characteristic of the cetacean brain is the very large and obviously well-developed neocortex. It is deeply fissured, twisted and folded intricately, and altogether strikingly similar to our own precious neocortex. In fact, it may be even more advanced. As the neocortex evolved, it covered the surface of more and more of the cerebral hemispheres. About one-third of the cerebrum of a hedgehog has a neocortical wrapping; in the rabbit, the neocortex covers just over half. The human cerebrum has neocortex over 95.9 percent of its surface. The common dolphin has an even more extensive covering: 97.8 percent. And the cerebrum in odontocetes appears even more elaborate than the human version. The human cerebrum has three distinct arching lobes in each hemisphere; the odontocete, four.

As the neocortex evolved from its reptilian beginnings to the mammals, it became more complex, adding distinct layers and differentiating, so that the cells in various sections are arranged differently. Thus, neurobiologists take the layering, or lamination, of the cortex and its specialization from region to region as indications of its complexity. By both criteria, the cetacean neocortex qualifies as highly evolved. Lamination and specialization in odontocetes compares to higher primates, including humans. Studies by Peter Morgane and Myron Jacobs show that the connections among neurons in the brain of a dolphin are as complex and intricate as those in primates. Also, when the connections in equivalent areas of the neocortex in humans or monkeys are put against those in dolphins, the results are strikingly similar. Study of what the various areas of the neocortex do, in either humans or cetaceans, is still in its infancy, but on the basis of their own work Morgane and Jacobs feel that much of the cortex is of the association type. Thus, by the current standards of neurobiology, the cetacean brain is a complex organ indicating complex function.

Yet it must also be remembered that the cetacean has a number of unique features that belie any quick comparison with the human brain. There is that massive cerebellum and the curious foreshortened architecture of the whole organ. The brain stem has a double bend unknown in any other mammalian order. Such uniqueness is hardly surprising. The cetaceans are the most divergent mammal group, one that has marched to the beat of a very different drummer for eons. The cetacean brain reached its present large size some 15 to 20 million years ago, while the human brain is a creation largely of the past 1.5 million years. Each brain evolved to meet the circumstances and demands of life. Given the strikingly different histories of humans and cetaceans in terms of both habitat and time, their brains certainly differ. But as yet we are ignorant of the differences beyond a cursory knowledge of brain anatomy.

One explanation for the size and complexity of the cetacean brain is that cetaceans, as air-breathing water-dwellers, must make many complex calculations about three-dimensional space in order to return to the surface to breathe. There is truth to this reasoning; it no doubt has something to do with the highly developed cerebellum of the cetacean brain; yet the argument is not sufficient. The sea is on the whole a considerably easier environment than the land. Temperatures vary little from day to night, and water is always available. Even air breathing in the sea poses less of a problem than it may appear to our terrestrial eyes, in that it has been solved by a creature of considerably less cerebral endowment than the cetaceans. In an evolutionary perspective, the elephant seal is just to the land side of a whale. Elephant seals come ashore in the winter to breed and calve, then return to the water in the spring and travel toward summer feeding grounds, where they live at sea for months on end. Thus, for part of the year, the elephant seal lives essentially as the whale does, feeding below and breathing above. In comparison to the cetaceans, however, the elephant seal is a cretin. In their fights for beachfront, the bulls crush newborn pups underfoot, despite the anguished screams of the infants. Parental care is almost nonexistent. To wean the young, the cows simply enter the water and swim away, leaving the pups on the beach, which the young

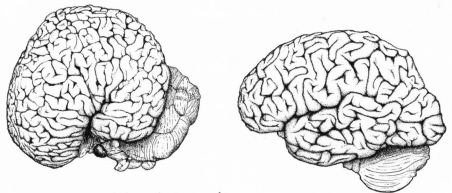

Dolphin brain and human brain to scale

ones themselves abandon when hunger forces them into the sea. Yet, however unappealing the elephant seal in human eyes, the animal adapts very well to life in the ocean. The northern species was thought extinct but for a relict population on Guadalupe Island around the turn of the century, but now the seal's population has recovered and the species is extending its range farther and farther north each year. The verdict: a mammal can live in the sea, breathe air, and have a run-of-the-mill brain all at the same time.

Another idea is that echolocation prompted the cetacean brain. This seems a better tack. As each mammalian order evolved it came to rely on a predominant sense, and that sense and the brain developed hand in hand. Carnivores smell and primates see; in simple terms, the one brain is olfactory and the other visual. Cetaceans are the virtuosos of sound, and hearing must have been central to the evolution of the cetacean brain. Yet, again, echolocation is not a sufficient explanation of the elaborate cetacean brain. One of the smallest-brained and most primitive cetaceans, the Amazon river dolphin, is an accomplished echolocator, able to hear up to 175 kHz and to distinguish by species fish of the same size.

A third possibility has to do with the social nature of the cetaceans. Darwin himself discussed what he called sexual selection; the evolution of a species is affected not only by the environment but also by the breeding patterns. In other words, how a species develops over time depends both on where it lives and how the ladies choose their lovers. Recently, theoretical biologists have been playing with fancier versions of Darwin's insight, putting the idea into mathematical terms and showing that social arrangement can profoundly affect the various parameters and rates that add up to evolutionary change. No doubt, human society, with its need for com-

Bowriding common dolphins (*Delphinus delphis*)

munication, had much to do with the evolution of our brains. Speech, society, and the neocortex are all of one evolutionary piece. Very possibly, a similar dynamic lies behind the complex brain of the cetaceans.

From what we know of cetacean society, it appears ordinary, and in captivity the animals accomplish little that cannot be explained with current theories of learning. The knee-jerk response of science is to write the animals off. Perhaps, though, the flaw is not the cetaceans, but the framework from which we look at them. Perhaps we have missed aspects of cetacean society and behavior because these aspects are unique to cetaceans. Possibly the complex cetacean brain has something to do with the physical and vocal mimicry. Possibly, too, cetaceans communicate in a mode unlike anything we have imagined.

Research into the language ability of the dolphin has assumed that that language, if it exists, must be like human language: sounds in units, which we call words, put into sequences, whose meaning depends on learned rules, which we call grammar. Dolphins, however, could have a wholly different way of getting their meaning across. "One very clever recent suggestion," writes Carl Sagan in *The Dragons of Eden*,". . . is that dolphin/dolphin communication involves a re-creation of the sonar reflection characteristics of the objects being described. In this view a dolphin does not 'say' a single word for shark, but rather transmits a set of clicks corresponding to the audio reflection spectrum it would obtain on irradiating a shark with sound waves in the dolphin's sound mode. The basic form of dolphin/dolphin communication in this view would be a sort of aural onomatopoeia, a drawing of audio frequency pictures — in this case, caricatures of a shark. We could well imagine the extension of such a language from concrete to abstract ideas. . . . It would be possible, then, for dolphins to create extraordinary audio images out of their imaginations rather than their experiences" (pp. 107–108).

We went to the dolphin brain seeking light, and have instead encountered only a thickening fog. Contemplate the cetacean brain, and the mist of mystery deepens all about. In the seas swim creatures with brains, the seat of the mind, hauntingly similar to our own, and we really have no idea what they are doing there.

There is more to the world than the science seeking to understand that world can explain. Science searches for order in a world that has an epicure's taste for chaos. To find meaning and pronounce laws, science lops off the odd corners and planes down the disconcerting bumps. We are ourselves one of the biggest bumps. The cetaceans are another.

John Lilly sensed this; he knew what it meant. He wanted to catch us by the short hair and force us to look at the cetaceans with open and wondering eyes. Then, somewhere and somehow, he abandoned the very standards of truth he most needed. His was a brave and true message. It deserved a better messenger.

Of Whalers and Whaling

*   *   *   *   *

*Canst thou draw out leviathan with
an hook? . . . Canst thou fill his skin
with barbed irons? or his head with fish
spears? Lay thine hand upon him, remember the battle, do no more.*
— J O B , 4 1 : 1 , 7 − 8

The Spear and the Iron

Cetaceans have been the prey of human hunters for millennia. Ancient middens are piled high with the bony remains of long-past feasts, and equally ancient art depicts the hunt that supplied those feasts. The earliest known drawing, from a cave on the Norwegian island of Røddøy and dating from about 2000 B.C., shows the crudely outlined figure of a hunter in a small boat chasing a pair of dolphins. Hunting dolphins in the shallows, though, is one thing. Seeking great whales at sea is quite another. Yet while the Røddøy islanders contented themselves with small game, Alaskan Eskimos were already setting upon Leviathan.

Killing small cetaceans differed little from taking other sea mammals. The Indians of the Northwest Coast, an area stretching from northern California to southeastern Alaska, hunted dolphins much as they sought sea lions, seals, and sea otters. Stalking his quarry from a fast slim-waisted canoe, the hunter struck with a barbed spear attached to a line. The hunters of some tribes held onto the line and played the struggling dolphin as an angler fights a fish, while others tied sealskin floats to the line and let the dolphin run, lest strong pressure on the line pull the spearhead through the dolphin's delicate skin. Once exhaustion and shock subdued the animal, the hunter brought it alongside, dispatched it with a club, and boated his prize. This basic method, spiced with all the variations of technology and custom that so delight anthropologists, typified dolphin hunting along many coasts.

Indians at Île aux Coudres on the St. Lawrence River set a trap for belugas, the white whales of the Arctic. The belugas habitually came upstream to fish on the incoming tide, then returned to sea on the ebb. Once the belugas had passed, the hunters built a weir of thin poles set three to ten feet apart across the whales' route back to sea. The thin poles vibrated in the ebbing tide, and the returning belugas, afraid to swim through what sounded to them like a solid wall, followed the line of poles into the shallows, where they were surrounded and killed.

The strong group-sense of some cetaceans, particularly the pilot whale, has often proved their undoing at human hands. Norse sea raiders long ago figured out how to take entire schools of pilot whales, and they carried the method to their colonies on the Hebrides, Orkney, Shetland, and Faeroe islands, where it is still practiced. Dozens of small boats form a semicircle about the school and direct it toward a gently shelving beach. Once the whales are in close, a hunter lances one. His intent is not to kill, but to wound. The injured whale runs ashore and beaches itself, and the other whales, which will not abandon the hurt animal as long as it lives, likewise ground themselves. At that, people fall upon the whales with lances and knives, and the slaughter rages.

In shallow water a single man with a lance can readily kill even a pilot whale or beluga, inoffensive animals that they are, but it usually takes all the effort and all the bravery of a whole crew of men to attack a great whale in the open sea. Aleut hunters, however, actually dared to set upon whales in pairs, using a technique found also in the Kuriles, Kamchatka, and Hokkaido and among the Chugach Eskimos. The hunters powdered the root of the poisonous plant aconite, mixed it with water and let it ferment, then coated their spears with the mixture. Now came the difficult part. In a two-man canoe the hunters stalked a whale, and the hunter in the rear maneuvered close enough for the forward hunter to drive a poisoned spear through the whale's blubber and into the muscles beneath. From then on, the success of the hunt was in the hands of fate. The aconite took several days to do its lethal work, and winds and currents had to be favorable to bring the carcass ashore. Not that the hunters spent their time idly waiting. They performed rituals, some of them involving human remains, intended to enlist the spirits of the dead in bringing the whale from the sea to their larders.

The Aleuts guarded the secret of aconite jealously, maintaining particular vigilance against the prying eyes of the white man. They even invented

Pilot whales (*Globicephala malaena*)

a story, told no doubt with an insouciance designed to instill real horror in their European listeners, that they concocted the whale-killer from the putrid fat of a rich man's corpse. The tale was told convincingly enough to take in the nineteenth-century anthropologists Pinart and Veniaminov, both of whom reported it as fact.

Poison whaling was at best a haphazard way to hunt, requiring a lucky coincidence of skill, good currents, and friendly spirits to deliver the prize to the hunter. More productive methods of primitive whaling made use of bigger boats, larger crews, and stronger gear, a combination of tools and men that provided the basic way of killing whales until this century.

Although all the tribes of the Northwest Coast took dolphins and, except for the finicky Tlingit, availed themselves of any dead whales that drifted ashore, only the tribes of southern Vancouver Island and the Olympic Peninsula actively hunted whales. The Nootka were the original and most avid whalers of the region. Tricks of the whaling trade, both practical and ritual, were handed down from father to son, and the man who led a boat crew and wielded the harpoon enjoyed considerable status and wealth.

Spring was the whaling season, the time when the grays migrated alongshore on their way to the Arctic. According to an account by John R. Jewitt, a seaman captured by Nootka warriors in a raid on his ship, a precise series of rituals preceded the whaling season, growing more and more intense as the first day of hunting neared. It was the duty and honor of the king to kill the first whale; no other harpooner was allowed to draw blood until the king had done so. Three times the king went into the woods to sing and pray for good whaling. After the third time he fasted for two days and wore a headband of bark and a spruce branch in his hair. In the final week, the king and his crew ate little, bathed several times a day, rubbed their skins raw with twigs and shells, and abstained from sex.

When the big day came, the crews of eight set out in their long canoes. Once they found a whale by sight or by the sound of its blows through the fog, the crew approached from the left and rear. The harpooner stood with one foot on the gunwhale and the other on the bow thwart, his weapon at the ready. That weapon was a fourteen- to eighteen-foot yew shaft on which was mounted a detachable harpoon made of a mussel shell cutting blade cemented between two elkhorn barbs with spruce gum. A sinew lanyard connected the harpoon to a long line of cedar withes with four sealskin floats painted in bright designs unique to the whaling crew. The harpoon shaft was too heavy to throw, and the paddlers nearly had to run the canoe onto the whale for the harpooner to strike his blow behind the left flipper. At the instant the harpoon went in, the harpooner ducked into the forward compartment and threw the line and its floats overboard while the crew sheered the canoe hard to port. The gray is very dangerous at close quarters, and the quick maneuver was necessary to avoid the flukes of the enraged whale.

The harpoon was a much more clever tool than its primitive makeup indicates. The sinew lanyard was secured to a hole in the center of the harpoon set at a right angle to the cutting blade. When the struck whale pulled against the lanyard, the force twisted the harpoon and toggled it against the blubber. The running whale dragged the long line behind, and the floats, bouncing and bright, marked the course of its flight and slowed its progress. A second canoe, the harpooner of which was usually a kinsman of the harpooner of the first boat, tried to cut off the fleeing whale and plant a second harpoon. Now the whale had to be pursued until it slowed enough for a harpooner to slash its fluke tendons with a spade. Exhausted and hamstrung, the whale was finished off with a bone-pointed lance driven in behind the left flipper again and again to pierce heart or lungs. The mouth of the dead whale was secured to keep it from shipping water, and the canoes towed the carcass to the village. Towing was the dull and backbreaking part of whaling and the reason why the men performed their prehunt rituals meticulously. Sloppiness irritated the gods, who punished offenders by sending struck whales seaward and leaving the men with a long haul home.

The Nootka probably lost five or ten whales for every one they killed. A harpoon could break when planted or during the chase; a line might part; the crew could strike a particularly combative whale and find themselves promptly capsized. In the whaling season Jewitt observed, the king lost one whale after another to harpoons that broke when he struck. Jewitt forged a new head from scrap iron, the king took a whale, and the white man lived to tell the tale.

Nootka whaling was quite similar in its tools and its cultural role to the whaling of the Inuit Eskimos of northern Alaska. Eskimo whale hunters enjoyed the same prestige as Nootka harpooners, and they likewise used a detachable toggle-head harpoon, a line with floats, lances, and an eight-man boat. That boat, the *umiak,* was made by covering a wooden frame with skins, preferably those of the bearded seal, to make it light and quick and quiet through the water. The Eskimos hunted and killed much as the Nootka did, harpooning the animal again and again and pursuing it to exhaustion and death on the tip of a lance.

The Eskimos, too, lost more whales than they took, only for them that loss was far more telling. The Nootka could come up empty-handed in whaling season and still eat well, for the sea and shore they inhabited abounded in seals, fish, and game. The Eskimos enjoyed no such luxury. They relied on the bowhead whale for half the food needed to last out the endless winter, and failure at whaling could well mean starvation. When Yankee whalers came to northern Alaska late in the 1800s, the Eskimos saw the superiority of their tools and they fast adopted them. Today, coastal Eskimos still fill their larders with bowhead meat taken in the traditional hunt with Yankee weapons.

The Eskimo hunt is a matter of waylaying the whales as they migrate. The bowheads of the western Arctic winter on the Siberian side of the Bering Sea. In spring they head north and east through the Bering Straits, following the shoreward lead in the ice pack around the northern Alaskan coast to their summer range in Mackenzie Bay. In the fall, with the ice edge much farther north than in the spring, the whales head almost due west to Wrangel Island before turning south to the Bering Straits and their wintering area. Hunters from villages along the bowheads' way hunt the whales as they pass. Most are killed in April and May, when the whales are confined to the lead and cannot escape to the open sea as they can in the fall.

At Point Hope and particularly at Barrow, the spring hunt is a central focus of village life, much as it was before the white man came to the

Arctic. As soon as the shoreward lead opens, the crews assemble them-selves and their equipment on the ice, where they will remain as long as the lead is open, the weather good, and the whales in migration. The men cut a trough in the ice and in it they lay the *umiak,* its bow extending out over the water for a quick launch. An iron harpoon lies in the bow, at-tached to a polyethylene line about 200 feet long. The line carries floats, either the inflated sealskins of antiquity or a red plastic model of the sort used to mark crab traps. Tents are pitched, stoves lighted against the cold, shotguns and rifles set out to kill any belugas or seals or waterfowl that happen by. But mostly the men settle down and watch the water for the spout of a bowhead. They wait and watch and watch and wait and watch some more.

And when a whale comes, they are in the *umiak* and on the water as if by reflex, moving quickly and silently. The springtime whales flee at the slightest noise, and the men must use all their considerable skill to paddle the boat in close enough for the harpooner to strike the whale just behind the head. As the iron harpoon sinks in, a long thin metal rod mounted parallel to it hits the whale's back and triggers a darting gun fastened to the harpoon shaft. The darting gun fires a pointed bomb-lance into the whale. Five seconds later, the bomb-lance explodes. Rarely, though, does the first bomb-lance kill the whale, and the wounded animal runs, trailing the line and floats.

When a whale is struck, all the boats on the water close in for the kill, since the first ten crews to reach the whale each receive a share. The whale is finished off with darting guns or shoulder guns, which are huge brass smoothbores that shoot a bomb-lance ten or twenty yards.

The carcass is towed to the edge of the shorefast ice and drawn out of the water by block and tackle to be butchered, a big task that keeps many hands busy for hours. The Eskimos use practically the whole whale: meat, baleen, gums, lips, flukes, flippers, brain, tongue, entrails, most viscera, and the earbones. When the butchering is finished, only a few entrails, odd bones, and the skull remain. Traditionally, the Inuit Eskimos returned the skull to the sea. According to their belief, the spirit of the whale survived, having merely lent its body to the people for their sustenance, and return-ing the skull sped the spirit on its search for a new home. At Point Hope this custom is still observed, but at Barrow the skull is often discarded on the ice and the spirit left to its own devices.

Whaling in Europe apparently originated among the Norsemen. They began in the usual small way, by closing the mouth of a fjord with a net and attacking the cetaceans trapped within. This method worked well against dolphins and pilot whales, which were small and slow enough to be cornered and speared, but the faster minke whales were another matter altogether. The Norsemen, who had the warrior's penchant for inventing new ways to kill, shot dirty crossbow bolts into the whales and waited for them to die of septicemia days later. Whale hunting was no sideline to the Norsemen. According to the saga *Laxdaela,* colonists were lured to Iceland by the abundance of whales in the waters roundabout. And the fatal inventiveness of the Norse hunters remained in evidence. For hunting the belugas and narwhals they discovered around Greenland, they fashioned a harpoon with movable barbs held in place by a metal ring, which was pushed back as the point speared the whale, springing the barbs and anchoring the harpoon in the flesh. Scandinavians settling along the Bay of Biscay in the tenth century A.D. lent this harpoon and presumably some of their skill as sea hunters to the neighboring Basques. That exchange opened a new chapter in the history of whaling.

The Bay of Biscay teemed with right whales in those days, and from salvaging drift carcasses, the Basques knew already the value of the animals. The borrowed Norse harpoon turned the Basques from scavenging to active hunting. The meat was salted, particularly the highly prized and high-priced tongue, and oil was rendered from the blubber. Baleen, too, proved a most useful material. After immersion in hot water, it could be molded and shaped to make riding crops, fishing rods, corset stays, knife handles, brushes, springs, chair seats, and so forth. Even the hairy fringe of the plates found a use as the helmet crests of armored knights.

By 1150 the town of San Sebastian was warehousing baleen for sale, and throughout the twelfth and thirteenth centuries more and more Basque towns entered the business of killing and marketing whales: Biarritz, Fuenterrabía, Guetaria, Bermeo, Castro-Urdiales, Lequeitio, Zarauz, Bayonne, and St. Jean de Luz.

Whaling in Europe prior to this period had been pretty much like subsistence whale-hunting anywhere: whales were killed to provide for the physical needs of a local population. Basque whaling offered a new wrinkle. The right whale changed from an animal killed to feed the locals into a source of raw materials for articles of trade. This was, and is, the very heart of commercial whaling. No longer was whaling regulated simply by the number of hungry mouths to be fed. Now it was subject to the complex vagaries of the marketplace. The implications of this change became clear only with time.

At first the Basques were shore whalers. A lookout on a high point scanned the sea for whales, and at his cry of a sighting the crews launched boats into the surf and made after their quarry with harpoon and lance.

In essence, this method differed little from Eskimo whaling, and the kill was equally low. According to records from Lequeitio from 1517 to 1661, one or two whales a year was average, with six a very good season indeed. However, despite this seemingly small take, the number of whales fell off fast, so much so that by 1660 rights were rare along the Biscay coast. Already the Basques were exploring new whaling grounds, enticed into their lumbering caravels and out of the sight of land by the big money to be made. Early in the seventeenth century baleen and oil sold for such high prices that one good-sized whale paid the costs of the voyage and turned every kill thereafter into pure, glittering profit. Pushing farther and farther afield as their skill grew, the first Basque ship reached the coast of Newfoundland in 1545. Within three decades some thirty to forty Basque whale ships were cruising the waters of what was then a spanking new New World.

The Basques proved to be innovators as well as explorers. They replaced the Norse harpoon with a fixed-barb model shaped like an arrowhead, which remained the basic whaleman's weapon until the mid-nineteenth century. They also tinkered with a new way of trying out blubber. Along the Bay of Biscay the blubber was boiled into oil in huge iron try-pots set on fires built on the beach. On the early short voyages, the blubber was packed into casks and tried out when the vessel returned to port. But the voyages of several months' duration that became the rule in the sixteenth century posed a problem. If the blubber was simply packed away, it turned into a fetid mess by the time the ship came home. Even if great care was taken to salt the blubber, it decayed enough that its oil burned poorly and fetched an equally poor price. Accordingly, the Basque whaling masters fiddled, with varying success, at building try-fires on board and boiling out the blubber at sea. Some masters proved better at this than others; the unlucky ones had their ships burned beneath them. But the Basques pioneered a technique that the Yankees later put to great use in the tropics.

The Basques had also pioneered something more sinister than a technological innovation or two. As yet, no case is known where subsistence whalers hunted their prey to extinction. There is nothing mystical in this fact, little that implies the "equilibrium" so beloved in contemporary environmentalist chic. A village, no matter how dearly its people relish whale meat, can eat only so much. Once all the hungry mouths and empty bellies are filled, there is no need to kill more. Hunters and prey can live side by side for millennia, as the Eskimos did before the Yankee whalers came. But the transformation of the whale into a natural resource like timber in a forest or coal in a seam, from which various products could be extracted for marketing far from the home grounds of either whales or whalers, changed the economics of the enterprise. No longer was the demand for dead whales limited by the number of mouths and bellies in the village. Now demand depended on the market fortunes of the products made from those whales. As long as baleen and oil fetched good prices, the logic of the marketplace made it "rational" — a word that, in an economic context, must always be used with a strong sense of the absurd — to keep killing whales until there weren't any more to kill. The Basques were the first "rational" whalers. They destroyed the right whales in their home waters and pushed on to a new source of supply in Newfoundland. Thus was established a pattern of commercial whaling repeated again and again to the very moment of this writing: locate a population of whales to exploit, kill them until they are too few to hunt further, push on to search out a new mother lode.

The next lode was located not by the Basques, however, but by the Dutch, a people whose nation just then was establishing itself as a commercial and maritime power. The Dutch were eager to reach the Orient to buy exotic goods and sell their own manufactures. Since the Spanish and Portuguese controlled the southern routes to Asia, Dutch explorers, enticed by their government's offer of a substantial fortune as a prize, headed north to search out a Northeast Passage. In the course of piloting such an expedition, Willem Barents discovered the island of Spitsbergen in 1596. Barents died on the ice of Novaya Zemlya the next winter without finding a new route to the Orient, but the expedition did report back home that the sea around Spitsbergen frothed with bowhead whales. The Dutch, ever quick to see a commercial possibility, prepared to go north and make the best of this one.

Their way was not unencumbered. The English, well aware of the fortune to be had from the whaling bonanza, maintained that Spitsbergen had in fact been discovered by Hugh Willoughby in 1553 and was therefore an English possession. The English hired Basque whalemen, provisioned ships, and headed north. The Basques, French, Germans, Danes, and Dutch followed close behind. At first the English were the most successful northern whalers, killing some 200 whales and taking home over 11,000 barrels of oil each season by 1617. But the presence of foreign whaling ships irritated the British, and they mixed their whaling with piracy. Soon the English were spending more time capturing ships, stealing cargoes, and destroying gear than whaling. The Dutch responded in kind by seizing and stripping five English whalers. The Dutch also proved to be superior hunters, becoming the most successful Spitsbergen whalers by the early 1620s. English pretensions to ownership of the island ended in 1625 when the ships of the financially troubled Muscovy Company arrived at the beginning of the season to find their camp and stored equipment burned by rival whalemen from York and Hull.

The early Spitsbergen whalemen worked from the shore in the manner of the Basques. Arriving in late spring, the whalemen set up a temporary camp of tents and tryworks on the beach. Whales were hunted from long-boats inside the island's many fjords. The animal was struck with one or two harpoons, left to run and pull the boat until it was exhausted, then finished off with repeated lance thrusts between the ribs. The carcass was towed to the shore camp, where workmen in an assembly line flensed off the blubber with long-handled knives and prepared it to be tried out. Baleen was cut out of the jaw, cleaned of gum tissue, and set out to dry. Except for a small amount of meat taken by the whalemen now and then to break the monotony of their diet, the rest of the carcass was discarded, creating such an abundance of carrion that crowds of seabirds and polar bears took up residence nearby. All summer long the whaling lasted. Then, in the fall, with the gales of winter close upon them, the whalemen took down their tents, cached their gear, filled the ships with the season's take, and headed home.

In 1623 the Dutch constructed a permanent town on the cold treeless shores of Spitsbergen and named it Smeerenburg (Blubber Town, or, perhaps, Fat City). Smeerenburg was more than the usual evil-smelling whaling station. It was a true Dutch village of shops, inns, and eating places where a whaleman could have a proper burgher's breakfast of hot rolls and coffee before a day on the flensing platform. The owners of the Noordsche company, which enjoyed a royal monopoly on whaling, could well afford to be generous. The take of oil and baleen piled so high by season's end that the whaling ships could not bring it all home, and extra cargo vessels had to be sent out from Holland in ballast to fetch the surplus.

But the pattern of exploitation, financial success, and extermination set by the Basques repeated itself with remarkable speed in Spitsbergen. By the late 1620s bowheads were so scarce inside the fjords that the whaling ships had to move out into open water and then, in slow steps of advancing decimation, north and east toward Franz Josef Land and Novaya Zemlya. Too far from the whaling grounds, Smeerenburg became a ghost town soon obliterated by the Arctic winter.

This first round of Spitsbergen whaling came to an end in the 1630s. The small French whaling fleet was captured by the Spanish navy, and, in retaliation for that and other indignities, the French army invaded Spain and burned the Basque ports. With the French and the Basques out of the picture, the Dutch had the north pretty much to themselves, until their attention was turned to the remainder of the Thirty Years' War and to repeated armed quarrels with the English. The second round of Spitsbergen whaling did not begin until the 1670s. The Noordsche company lost the protection of its monopoly, and in the burst of free enterprise that followed, competing Dutch companies had 260 ships manned by 14,000 sailors working the Arctic by 1680. Whaling ships from other countries also hunted those waters, but the Dutch were the most numerous and the most productive. Of almost 2,000 whales killed in 1698, nearly 1,300 were taken by whaling ships from Holland. The Dutch worked ever farther and farther west in pursuit of whales, edging around the tip of Greenland and entering the Davis Strait into Baffin Bay. In the period between 1675 and 1721 Dutch thoroughness killed 32,907 whales.

This big kill brought the Dutch big money. In the opening decade of the eighteenth century, the Dutch took 8,537 whales, sold their baleen and oil for 26,000,000 florins, and kept 4,750,000 florins as net income. Translating money from one currency to another over the reach of almost three centuries, with their many inflations and economic differences, is a tricky business, but even with the approximate figures these qualifications permit, it is apparent that the Dutch did very well. Based on conversion rates worked out by historian Will Durant, the florin of the early 1700s equaled roughly $11 today. Thus every whale the Dutch killed in those years brought in about $33,500, and the decade of whaling earned $286,000,-000 for a net income of $52,250,000. These income figures compare almost exactly to the earnings of Norwegian whaling companies in the late 1950s and early 1960s. But the Norwegians were using factory ships several hundred feet long and fleets of catcher boats with radar and harpoon cannons, while the Dutch whalemen dared the polar seas in small dull sailers, rowed their longboats by hand, and killed with weapons of poorly forged iron.

Yet this labor-intensive, low-capital technology swept the northern whaling grounds clean. Catches averaged close to 2,000 yearly around 1690, fell to about 800 by 1720. The remaining whales were to be found deeper and deeper in the ice, where the already risky activity of whaling became perilous in the extreme. Life and limb were lost regularly, not so much to the inoffensive bowheads as to the caprice of Arctic climate and sea. At any time the ice could close and trap a ship, crushing it slowly or holding it fast throughout the long and dark winter. The Dutch record relatively few mishaps, a tribute to their seamanship and prudence. But the English, again driven to overtake their competitors in the Netherlands and proceeding deeper into the Arctic as bowheads became scarcer, lost ships and crews almost every season. Shipowners added to these perils, for business reasons, of course. Whaling ships were provisioned for a six months' voyage only, so that any ship caught in the closing ice at the end of summer faced the winter with most of its food already gone. Scurvy or starvation finished what the ice had started. Despite repeated losses of men, owners refused to bear the extra cost of provisioning their ships for a full year as insurance against disaster.

The handwriting was on the wall: whales were ever scarcer, risks ever greater, earnings ever smaller. The Dutch began withdrawing from the whaling business, sending out fewer ships each season. Their departure was hastened when in 1798 an English convoy force destroyed the Dutch whaling fleet. Now the British held in the north the predominance they had long desired. Their accession, though, was anticlimactic. The heyday of whaling in the eastern Arctic was already the better part of a century behind them.

In Japan, whaling went through an independent but identical change from subsistence to commerce. For centuries whales had been hunted for food in the islands. Then in 1606, during one of Japan's many civil wars, the defeated army of the Kamakura shogunate dug in at Taizi in central Japan and took up whaling to feed itself and to sell the leftovers for cash. This was the beginning of commercial whaling in Japan. Later in that same century whaling underwent a revolution of method, when whalemen on Kyushu perfected a way of killing and securing whales with nets.

From a high point a lookout watched the sea for whales entering shallow water, and at his alarm, a crowd of boats carrying five or six men apiece set out. The hunter boats maneuvered the whale away from the open sea and herded it toward shore, where three pairs of net boats waited. Each pair of net boats carried an outfit comprising 38 long but shallow nets joined one to another by light lines. At a signal each net boat in a pair rowed away from the other and strung the outfit between them and across the whale's path. The hunters drove the whale into the nets. The whale, thrashing in panic, broke the lines joining the nets and entangled itself even more. When the snared whale surfaced, a harpooner planted an iron or two in it to impede its flight further. Once exhausted by the struggle, the whale was lanced again and again, a hobbled Gulliver needled by Lilliputians, until nearly dead. Then a single harpooner, displaying a samurai's measure of bravery, dove into the water and cut a hole in the whale's snout, to which a securing line was attached. More harpooners swam under the whale, towing lines behind them to hold the body up in case the animal sank as it died. Thus supported and secured, the carcass was towed ashore.

The Japanese were skilled. They captured rights and humpbacks and even occasional finbacks, which European whalemen found far too fast to pursue. With gray whales, however, the Japanese quickly learned to put their nets aside and use only harpoons. Entrapment simply enraged the gray and made an animal always dangerous at close quarters into something murderous.

The Japanese whalemen processed their catch in an assembly-line fashion easily equal to Dutch efficiency at Smeerenburg, but, where the Europeans took only the blubber and baleen and squanderously discarded the rest, the Japanese made use of nearly the whole animal. Bones as well as blubber were boiled for oil, and the remains of the bones were ground into meal for fertilizer. The meat and·flukes were salted and cured. Intestines and viscera were likewise made into food. Tendons were dissected out of the meat and dried to make ropes and lashings.

Net whaling continued into the twentieth century when, like much Japanese tradition, it fell by the wayside of the industrial revolution. The last whale caught in nets was killed off the coast of Yamaguchi prefecture on southwestern Honshu in 1909.

The settlers of New England were a God-fearing and money-making folk, equipped by temperament and theology to convert the abundant resources of the new land into calculable wealth. Not the least of these resources was the whale. Right whales and evidently a small number of grays frequented the shoreward waters from Long Island to Massachusetts Bay. Beginning with drift whaling, the Yankees graduated to hunting close to shore by the early 1650s.

The people of Nantucket, the Quaker island whose name was soon synonymous with whaling, got into the enterprise almost by accident. A "scrag" whale, a small right or gray, crossed the sandbar into Nantucket harbor and, apparently seeking refuge because of injury or disease, remained there three days. That was enough time for the blacksmith to fashion a harpoon, with which the islanders attacked the distressed whale and killed it. Their appetite for whaling whetted, the Nantucketers made an agreement with Ichabod Paddock of Cape Cod to teach them how to kill whales and try the blubber. Paddock arrived on the island in 1690, and the Nantucketers learned his lessons so well that the history of American whaling in the 1700s is essentially the history of Nantucket.

The Yankees, themselves transplanted Europeans, used European methods of whaling, but they did learn a trick or two from the local Indians. When Indians hunted whales, they fastened logs onto the line to act as drags and slow and tire the struck whale. The Yankees copied the idea, fashioning any number of tublike sea anchors, called drogues. They also borrowed Indians as whalemen, using them as lookouts and as pullers in the whaleboats.

The Nantucketers knew there were other whales in the sea besides the rights and grays they hunted alongshore. The drift carcass of a sperm whale had once washed up on a local beach, but since the lookouts only very rarely spotted the distinctive oblique blow of the sperm whale, the Nantucketers thought this species highly uncommon. Then in 1712, says a dubious legend, an untoward wind carried a whaleboat commanded by Christopher Hussey out to sea. There upon the deep, tossed by storm swells, Hussey and his companions found themselves in the very midst of a pod of sperm whales. A whaleman always, Hussey killed one of the sperms and managed to get his boat, crew, and catch back to Nantucket. Unlike the right, gray, and humpback, all of which spend much of their lives within sight of land, the sperm whale is a true deep-water cetacean usually found well offshore. The sperm whale wasn't rare at all; the Nantucketers simply hadn't known where to look. Once they knew, they looked hard. The men left the island on sloops of 40 to 50 feet in length that carried a crew of 15 and two whaleboats and stayed at sea about two weeks.

The right whale and the sperm yielded different products. The right provided baleen and whale oil, which was used primarily as an illuminant and in industrial processes like the manufacture of steel and jute. A good-sized right whale yielded about 100 barrels of oil (at 31.5 gallons to the barrel). Few sperm whales ever produced so much, but sperm oil was a superior illuminant that fetched two or three times the price of whale oil. Even more valuable was the white wax of the spermaceti organ. It was mixed with the "junk" fat of the upper jaw to make "head-matter" and fashioned into the highest-quality, brightest-burning candles then available. Sick sperm whales also occasionally rewarded their killers with ambergris, a lumpy waxlike material formed in the intestines by the concretion of bile salts. When first removed from the guts, ambergris smells all too strongly of its origins, but it soon takes on a pleasant earthy aroma. Ambergris was highly prized as a fixative for perfumes and volatile essences. The sperm whale, being a toothed cetacean, of course gave no baleen, but there was a small market for the "ivory" of its teeth and lower jawbone.

Step by step the whaling sloops of Nantucket extended their voyages from the shoals about the island to Georges Bank and out into the sea river of the Gulf Stream. Soon the Nantucketers were hunting the Davis Strait and Baffin Bay. By 1748 Nantucket had sixty whalers at sea and were taking an annual catch worth about $100,000.

The Yankees proved their commercial mettle despite concerted odds. English whalers earned a bounty just for outfitting, and they brought home their oil duty-free. The Yankees earned no bounty and paid a duty for oil exported to England, yet they turned a profit, while English firms lost money to the tune of £178,000 in eight years. England's own whalers could not supply all the whale and sperm oil the nation needed, and the bounty and duty were dropped in favor of importing more Yankee oil. But the English still insisted on running things their way. Colonial whalemen were prohibited against trading with any nation other than England, and when conquest opened the Gulf of St. Lawrence and the Straits of Belle Isle to whaling, Canadian governors imposed on the Yankees restrictions so onerous that most gave up whaling in those waters. The Nantucketers turned south, looking for new whaling grounds. Already the New Englanders had scouted the sperm grounds around the Azores and the Cape Verde Islands. They continued along the coast of Africa to Walvis Bay in Namibia (South-West Africa), hunted the Brazil Banks off South America, and extended their voyages deep into the South Atlantic to the Falkland Islands in 1774.

The whaling industry grew as fast as it extended its range. Between 1760 and 1775 the American whaling fleet more than quadrupled, from about 70 vessels to just over 400. Nantucket harbored 150 of these ships, and the remainder set out from Dartmouth, Lynn, Martha's Vineyard, Boston, and other New England ports. The whaling vessel had grown in size, too. The Nantucket vessels that headed south for sperms averaged 120 tons and stayed at sea as long as 12 months. The effort and the investment paid off. In 1774 whaling brought Nantucket $500,000, an increase in income of more than five times in less than thirty years.

The Nantucketers whaled with an insular singlemindedness that blinkered them to politics, and the Revolutionary War burst upon them unaware. As Quakers, the Nantucketers took neither side, but like all pacifists in time of war, they found themselves between a rock and a hard place. Loyalists seeking escape from the Massachusetts firebrands came to Nantucket, giving rebel privateers cause to blockade the harbor and seize ships as they returned from the South Atlantic ignorant of the hostilities. Yet, despite the abuse the Nantucketers suffered from the Continentals, the English slated the island for the destructive attentions of a punitive expedition that also struck at New Bedford, Falmouth, Wareham, and Martha's Vineyard. Fortunately, a gale drove the expedition back to New York. But Tory privateers picked up where the Royal Navy left off and raided the island, stealing what they could and burning the rest. The English also tried to turn a profit from Nantucket's persecution. The Royal Navy took no fewer than 134 ships, many whalers among them, and killed or captured 1,200 seamen. The prisoners were given the "choice" of serving out the war's duration on whalers or men-o'-war. Thus throughout the revolution English lamps burned oil taken by Nantucketer prisoners of war. For some bizarre constellation of reasons, the whaling business in England, despite all the maritime skill of that island nation, succeeded only when it relied on naval muscle or the press gang.

As soon as the war ended, Nantucket and the other whaling ports returned to their enterprise with such a will that they promptly glutted the American market. Nantucket, for which whaling was the economic mainstay, fell upon the hardest of hard times. A group of Nantucketers led by William Rotch left the island and set up a base of operations in Dunkirk, France. Yet another group emigrated first to Nova Scotia and then, at the invitation of the British government, to Milford Haven in Wales.

England had the largest whaling fleet in the world at this time, over 300 ships, but the industry relied heavily on outside talent. While Englishmen commanded the Arctic whalers, Nantucketers mastered and crewed the southern sperm-whalers. In 1789 the *Emilia,* an English vessel owned by Enderby & Sons, Ltd., became the first whaler to round the Horn. The vessel's master hailed from Nantucket, and his mate, Archaelus Hammond, also a Nantucketer, killed the first Pacific sperm whale off the coast of Chile. The *Emilia* had discovered an abundance of sperm whales at a time when Rotch's market in France, a new export business in spermaceti candles, and a switch from tallow tapers to whale oil in streetlamps raised the prices for whale products. In 1791 six American whalers were outfitted for the Pacific, opening the classic period of South Seas whaling later depicted in *Moby Dick.*

But politics continued to divert the Yankees from their business of killing whales. The French Revolution and war between France and England eventually closed Rotch's Dunkirk operation. British and Dutch raiders and French men-o'-war seized American whaling vessels for various reasons and pretexts. Yet, from Nantucket's point of view, the upheaval was not altogether bad, for it brought many of the expatriate islanders home. They returned to their ways of searching new waters for whales, hunting up the South American coast to Peru and the Galápagos and rounding Good Hope into the Indian Ocean. Behind the Nantucketers came the whalemen of New Bedford, Falmouth, New London, Sag Harbor, Greenwich, and Westport. The War of 1812 caught most of the New England whalers at sea, where they fell easy and unknowing prey to British privateers, armed whalers, and the Royal Navy. To add to the peril, the Peruvians seized ships of either belligerent, but since they considered the United

States the weaker of the two nations, they preferred American prizes. By the end of the war, few Yankee whalers in the Pacific remained afloat or free.

But, for the first time, the Yankees had the last warlike laugh. Captain David Porter sailed the United States cruiser *Essex* into the Pacific, and by derring-do and deception put a Peruvian privateer and over a dozen armed English whalers out of commission before his ship was shot up and forced to surrender.

The end of the war ushered in an era of unprecedented growth for the whaling industry and of unrelieved destruction for the whales. By 1822 the Nantucket fleet had grown from about two dozen vessels to 75 ships and 7 brigs. The Greenland–Davis Strait grounds that the English relied on had been hunted into decline, and that fall-off in the supply of whale oil made for good market prices. More and more venture capital entered the industry, building bigger ships and outfitting them for three-year voyages into the distant Pacific whaling grounds. By the mid-1820s Yankee whaling had more than made up past losses to war. The boom was on.

Whaling fever swept the East. Practically every port from Edenton, North Carolina, to Bucksport, Maine, fitted out a whaler or two and sent it to sea. In 1835 the Yankee whaling fleet numbered an even 500 vessels, which averaged 290 tons and which brought in oil and baleen worth over $6 million. Depression and increased operating expenses knocked most of the new whaling ports out of the trade in the 1840s, but the fleet continued to grow, particularly in New Bedford. The year 1846 saw the largest number of whalers ever to sail from American ports: 736 ships, barks, brigs, schooners, and sloops. About one-third of these vessels took two- and three-year sperming voyages into the Pacific and Indian oceans, over half were in right-whaling, and the remainder made shorter excursions to the Atlantic sperming grounds. The cargoes brought in were as huge as the effort expended to secure them. Whaling vessels returning to port in 1847 came with 433,903 barrels of sperm and whale oil. Baleen cargoes hit their high point in 1853 with a take of 5,652,300 pounds. The Yankees killed 8,000 to 10,000 whales a year and made annual earnings that topped $10 million five times in the 1850s.

During this period New Bedford overtook Nantucket as the principal Yankee whaling port. In fact, Nantucket, the pioneer of long-distance sperm-whaling, finally left the business altogether. Nantucket's troubles lay in her own harbor. In the early days of shallow-draft sloops and schooners, the sandbar across the harbor mouth posed little problem, but as voyages lengthened and ships grew larger and drew deeper, the bar became a real obstacle. A fully loaded long-distance sperm-whaler could not cross safely and had either to off-load into whaleboats back of the bar or dock instead at Martha's Vineyard. This complicated matters and increased costs considerably. The Nantucketers fashioned floating dry-docks to bring ships across the bar and petitioned the federal government repeatedly to build a breakwater. Then a fire destroyed Nantucket's business district in 1846, and the California gold rush drew away most of the island's young men. The government didn't get around to building the breakwater until 1881, when it was something of an afterthought. The last whaler had sailed from Nantucket twelve years earlier.

New Bedford led American whaling and America led the world. Other nations did whale, most notably England, but of all the whaling vessels in the world in the 1840s, three-fourths were American. The lamps and candles that lighted the mansions, houses, and hovels of New World and Old came from right and sperm whales killed by Yankee whalemen.

It is a commonplace now to think of Yankee whaling as a cottage industry that kept a few old salts busy while the rest of the nation busied itself inland with the real tasks of clearing forests, building railroads and factories, and shooting Indians. In fact, whaling was the third largest industry in industrial Massachusetts, surpassed only by textiles and shoemaking. Whaling was more than 600 or 700 ships and their 17,500 officers and men. The industry put people to work at infitting and outfitting, processing oil and baleen, making spermaceti candles, and building and maintaining ships. E. P. Hohman, whose early study of the economics of Yankee whaling remains the best, estimated that all these related enterprises employed 70,000 people — at a time when the population numbered fewer than 30 million — and created about $70 million in income — at a time when the gross national product had yet to top $10 billion. Whaling was big business.

The whalemen went about their business at sea with considerable purposefulness and exactness. Whales are found in particular areas of the oceans at given times of the year, the precise location determined principally by the availability of that species's food and by its calving and mating habits. The whalemen understood little of the whys and wherefores of the whales' migrations, nor did they care to know. Their business was killing whales, and the way to accomplish that was to be on the whales' sea haunts, or grounds, when the whales were. The masters plotted their courses to take them to the right place at the right time, choosing the spot and the season from past experience. A ship departing New England in early summer would work the Azores sperm grounds until October, spend the southern summer in the Brazil Banks hunting for both sperm and right, round the Horn and provision in Chile or Peru in the spring, then cruise

Atlantic right whale (*Balaena glacialis*)

the Onshore Grounds along the South American coast for sperms through
the winter. A ship leaving in autumn was more likely to go by way of the
Cape of Good Hope, stopping on the way to kill humpbacks in the Gulf
of Guinea and hunting rights at the Cape, in the Mozambique Channel,
and in the Indian Ocean. Southern autumn found the Good Hope whaler near
New Zealand and Australia, where rights and sperms and humpbacks could
be had, and winter would likely draw the vessel toward the sperm grounds
farther north.

Once on the grounds, the Yankees combed the sea methodically. The
ship sailed slowly across, a lookout in each masthead from dawn to dusk.
After night fell, sail was brought in and the ship stayed where it was, to
pick up the hunt in the morning precisely where it had been left off the
evening before. When the ship reached the edge of the grounds, it came
about and crossed again on a parallel course, repeating the crossings until
the whole grounds had been covered. The Yankees were equally thorough
in their killing. They took any and every whale they could secure, be it
bull or cow or calf, and they hunted the same reliable and well-known
area until too few whales remained to be worth continued hunting. The
whalemen were always on the lookout for new places to hunt, their minds
on oil and baleen and dollars.

In the opening days of the golden age of Yankee whaling, most of the
sperms came from the Atlantic and the Onshore Grounds. Right whales,
already rare along the New England and Canadian coasts, were hunted in
the South Atlantic, the southern Indian Ocean, and around Australia and
New Zealand. In 1818 the Offshore Grounds, a sperm-whale area a thousand miles west of Peru, were found. The Japan Grounds, east of Japan
out to the Bonin Islands, were first hunted in 1819 and 1820 by whalemen
who went there because of a merchant captain's report of large sperm
whales in the area. By 1838 whalemen were working the Northwest Coast,

whose Alaskan and Canadian reaches were abundant with right whales. Kamchatka and the Sea of Okhotsk, good grounds for bowheads, were added to the whaling itinerary soon thereafter. In 1848 a Sag Harbor whaler went through the Bering Straits and killed the first Arctic Ocean bowhead.

There was good reason to the Yankees' urge northward: necessity. Wherever the Yankees had been, few whales remained. Methodical hunting had turned the grounds of the Atlantic and the South Seas into little more than memories. Whales survived there, to be sure, but they were scarce and scattered over too much ocean to hunt profitably. The marketplace demanded that whalemen kill whales, and the whalemen responded by pushing into unknown and as yet unhunted waters. The Yankees fanned out across the Pacific, searching every nook and cranny and back bay in all that world of water.

The vessel the Yankees went in was a small ship or bark of around 350 tons, about 100 feet long and 28 feet in the beam, built bulky to hold the great mass of whaling gear and oil casks and to withstand the fearsome stress of a whale lashed to its side in heavy seas. Ungainly but strong, whalers were rarely lost to storms or collisions.

Aloft a whaler was rigged much like any vessel of the era, except for hoops in the mastheads where the lookouts stood their watches and scanned the sea for whales. When whales were sighted, the captain steered his ship into as advantageous a position as wind and whales allowed and ordered the boats away. Usually the captain himself stayed aboard, often going aloft to guide the boats to the whales by signals.

The typical whaling ship carried four whaleboats mounted on davits: a single starboard boat for the captain or fourth mate; larboard, waist, and bow boats for the first, second, and third mates. The Yankee whaleboat was a paragon of lightness and speed, a clinker-built craft made of white cedar with its sides widely flared and both ends pointed. Though only 30 feet long, the whaleboat carried a crew of six plus all the gear needed to chase, kill, and tow the whale. Harpoons and lances waited in the bow, their points snugged safely inside wooden sheaths. Two tubs of line, previously cleared of kinks and twists by towing astern and by coiling first one way and then the other again and again, lay in the stern. Only after the whaleboat had been lowered into the water was the line attached to the harpoons. It was brought up through a hole in the decked-over stern, run around a stout post called the loggerhead, then taken forward over the oars to a groove in the bow called the chocks. A slender chock-pin of wood or baleen held the line in the chocks. The chock-pin was purposely flimsy, so that any kink forming in the running line and catching against the chocks would break the pin and let the line run free, saving the boat from being dragged under by a sounding whale.

With the boat in the water and the harpoons bent onto the line, the mate took his position in the stern and steered the whaleboat in pursuit. He brought the boat directly onto the whale's head or flukes to stay out of the animal's field of vision, and he avoided the slick, or "glip," left by the whale's tail as it swam, believing that touching it would alarm the whale. The men at the oars pulled with their backs to the whale. Perhaps this was for the best. The harpoon could be cast only a few yards, and the closer the boat came, the harder and deeper the harpooner could strike. At such a short distance a whale is fearfully huge, but the oarsmen had no occasion to contemplate the whale's size, pulling blind until the boat was nearly on top of it. At the mate's command, the harpooner shipped his oar, turned around and leaned against the forward thwart to steady himself, and raised the harpoon high by its wooden shaft. At the right moment he darted the iron into the whale.

The moment was right when the back was arched and the skin tight. Should the harpooner throw when the skin was slack, the wound could open as the whale bent to dive and the harpoon would pull free. The invention of the toggle-iron in 1848 by James Temple, a freedman blacksmith of New Bedford, solved this problem somewhat. When the line pulled on the toggle-iron, the tip flipped open on a pivot and anchored itself against the backside of the blubber. The toggle-iron was a distinct improvement on the arrowhead-shaped harpoon and soon replaced it.

Many whaling crews tied two harpoons to the line, and the harpooner had to plant the second iron or throw it overboard, lest it fly around the boat and impale an oarsman. As soon as the whale was fast and the second iron cleared, the mate and the harpooner performed a ritual that, although an irrational risk of life and limb, was an unwavering Yankee custom. While the whale fled and line flew out so fast that it smoked on the loggerhead, the mate and harpooner actually changed places. The one went to the stern to control the line on the loggerhead and to steer — the reason why the trade called harpooners "boatsteerers" — while the mate advanced to the bow — earning thus the name "boat header."

The struck whale usually either sounded deeply or ran across the surface, pulling the boat behind. When it rose or slowed, the oarsmen drew in line and warped the boat close enough for the mate in the bow to go to work with his lance, a long barbless spear attached to a lanyard. The mate cast the lance into the whale, aiming behind the left flipper, and retrieved it with the lanyard. Once the whale was in a bad enough way for the boat to come in very close, the mate thrust the lance between the ribs and churned it to pierce and tear the lungs. This was a brutal business, a giant death inflicted inchwise, the whale finally spouting blood and expiring in a grotesque flurry that scummed the sea with gore and vomit.

It commonly took one to six hours to kill a whale, and some chases lasted as long as forty hours. But with the whale finally dead, the crew had only begun to work. Unless the ship lay to windward and could come alongside to pick up the kill, the carcass had to be towed back by oar and muscle. The dead whale was made fast to the ship's starboard side, its tail to the bow. A removable section of the side amidships, called the gangway, was taken out, and a scaffold, or cutting-stage, was lowered and swung into place over the carcass. The scuppers were plugged to keep everything precious on board, and two blocks and tackles were sent aloft and secured to the mainmast. The mates took their places on the cutting-stage and set to work with long-handled knives and sharp spades.

Cutting in was every bit as dangerous as the chase and kill. Swinging gear and blubber pieces, knives and spades, a pitching deck, and hordes of sharks attracted to the carcass could maim or kill and often did.

To cut in a sperm whale, a hole was chopped in the blubber between eye and fin, and a blubber hook on the block and tackle snugged into it. With this purchase, the carcass could be rolled on its back. The lower jaw was cut and wrenched from its socket and hoisted on deck. Then, by complicated maneuverings of the securing chains and tackles and by chopping with the spades, the head was broken free, the upper jaw separated and discarded, and the junk and case hauled astern and made fast. Next the carcass was stripped of blubber like a banana being skinned in a spiral. This long blanket piece was cut into smaller sections and lowered into a blubber room below decks. If the whale looked sickly, the mates plumbed the guts with their spades and sniffed them for a telltale whiff of ambergris before setting the carcass adrift. The head was processed last. A small head could be hoisted aboard and the spermaceti bailed from the case on deck. A big head, though, would break the tackle or snap the mast. A crewman wearing a safety belt went overboard, and while men armed with cutting spades killed any curious sharks, he made the lacings necessary to lift the rear of the head even with the gangway, where the case could be tapped and bailed. The layer of junk was cut free and brought aboard, and the empty head was set adrift.

When a right whale had been brought in, the upper jaw was broken free and brought on deck. The baleen was cut from the jaw in sections and stowed below, later to be scraped, washed of gum tissue, and dried in the sun. The lower jaw was also hoisted aboard, for the lips and tongue contained considerable oil. Then the blubber was stripped off in a scarf much as with a sperm whale, except that a fin chain and toggles were used in place of a blubber hook.

As the carcass was being stripped, a fire was set in the brick tryworks, which sat in a foot-deep pan of seawater on the deck forward of the fore hatch. With big two-handled knives, crewmen cut and minced the blubber into smaller pieces, which were dropped into big iron pots on the tryworks to be boiled. The harpooners ladled the oil into a cooler, from which it was piped into casks and stowed in the hold. The leftovers of the tried-out blubber were fed to the fire as fuel.

Cutting in and trying out continued round the clock, the watches alternating six-hour shifts, until the last cask was in the hold. If several whales had been killed, the work could go on nonstop for three or four days. By that time the entire ship and all the men were filthy with blood, oil, and soot. Lye was made from the ashes in the tryworks, and the crew cleaned the ship and themselves. Then the men could rest until the next whale was sighted. Weeks, even months might pass. But it could be that even as the men fell into their bunks, the lookout raised a pod, and the whole wearying round began again.

All the many tasks and chores involved in whaling necessitated large crews of 25 to 35 men. A typical ship's company comprised the captain, three or four mates, three harpooners, cooper, blacksmith, cook, steward, possibly a carpenter, and the rest seamen, green hands, and boys. The whaling ship was unique in that it had three castes rather than the usual two. The captain and mates lived in the cabin in the stern and ate the best food. The boatsteerers and cooper occupied steerage amidships, and although they ate not quite so well as the officers, their meals were served at the cabin table after the captain and mates had eaten. Boatsteerers were something like warrant officers, above the men and below the masters, and until a harpooner led the infamous *Globe* mutiny of 1824, boatsteerers even stood night watches. The men lived in the forecastle and ate from plates balanced on their laps. The poor setting fitted the food, which was a monotonous reserving of salt beef, hard biscuits, flour, dried peas and beans, and rice, unrelieved by the soft bread, butter, and sugar the officers enjoyed. When the men finished eating, they put plates and forks into the netting over their bunks. Cockroaches ate them clean before the next meal. The whalemen didn't really mind the cockroaches; they believed the vermin ate bedbugs, which were much more troublesome.

Pests were only one drawback to life in the forecastle. The room itself was terribly cramped, twelve to sixteen feet square with the butt of the foremast occupying the center of the floor. The bunks, sixteen or eighteen of them, were built into the sides. Light and air entered only from the hatch, which was also the way in or out, not only for the men, but also for every sea the whaler shipped.

Such hardship could be justified if a whaling hand made good money for his trouble, but compensation for the average seaman was abysmal. All members of the crew, from captain to cabin boy, were paid by the "lay" system. This arrangement, in which each man earned an agreed share of the cargo, can be traced back to Dutch and French ships in Spitsbergen. Among the Nantucketers, the system of shares made eminent good sense. Master and men all came from the same small community, all worked together, all shared in the risks and rewards of their labor. But the system remained in use after whaling had expanded beyond the Nantucket family enterprise, and it changed from communal economics to another way of squeezing the worker.

Captains, mates, coopers, and boatsteerers earned lays of from $1/8$ to $1/100$; $1/100$ to $1/160$ was paid to seamen, stewards, cooks and blacksmiths; greenhands and boys worked for $1/160$ to as little as $1/350$. When time came for settling up a few days after the ship returned to port, the accountant figured the current market value of the cargo and computed each man's share. This was not, however, the amount the crewman got, for there were certain deductions made against earnings: $75 for clothing and gear provided at the beginning of the voyage, cash advances made in ports of call, charges for the medicine chest, charges for outfitting the ship and unloading the cargo, the price of all clothes and tobacco purchased from the ship's store ("slop chest"), even expenses for recapturing the man if he jumped ship. To this already lengthy list of debits was added flat interest of 20 to 30 percent, with usurious rates of as much as 60 percent known after 1850. The accountant subtracted the charges and interest from the fractional share of the cargo and paid the man the remainder. Now and again a forecastle hand who had spent the last four years at sea actually found himself owing the ship money.

Even as the Yankee industry prospered, the earnings of crewmen actually decreased. In 1795 the whaling ship *Lydia* paid $1/45$ to a boatsteerer and $1/60$ to a seaman. In 1847 the same crewmen on the *Brighton* earned $1/75$ and $1/160$. According to figures worked out by Hohman, a forecastle hand in 1850 earned about 20 cents a day, compared to 40 to 60 cents a day for unskilled labor ashore. Nor was this simply a land-sea difference. Merchant seamen earned three times as much as whalemen.

Such dismal financial prospects contributed to what is often called the "deterioration" of whaling crews. The old New England hands and boys from the back country eager to trade farm for sea were replaced by Portuguese from the Azores, Polynesians, and Cape Verde blacks.

Numerous abuses arose. Crews were filled out by agents paid per head, and such landsharks, as the hands unaffectionately called them, resorted regularly to overdoses of rum, shylocked debts, or con games with prostitutes to fill the forecastle with enough warm bodies for the vessel to clear port. Since each man had $75 of clothing and bedding paid for by the ship's owners, the officers made sure that no one deserted in the early days of the voyage, but in the last months some actually provoked desertions, thus appropriating the earnings due the missing man. The captain got a percentage on all goods sold from the slop chest, and, like a store owner in the ghetto, he sold inferior goods to a captive audience at inflated prices.

Behind such a system of sanctioned thievery, a proper cynic would expect to find a conspiracy of robber-baron capitalists, all gathered in some corporate boardroom to chew fat cigars and plot further outrages against nature and labor. Curiously, the concentration of economic power that characterized railroads, for example, did not hold true for whaling.

Whaling as a whole was not wildly profitable. Charles Enderby of the English whaling firm of the same name pegged the average annual profit of sperm whaling at 1.3 percent and of right whaling at 6.5 percent. There were, however, few average voyages. The industry was a crazy quilt of bonanza and ruin. At one extreme was the brig *Emeline,* which spent two years and two months whaling, lost its captain in the process, and came back to New Bedford with but 10 barrels of oil. At the other extreme was the *Lagoda,* also of New Bedford, which between 1843 and 1973 made eight voyages at an average profit of 157 percent, its lowest return being 67 percent and the highest 364 percent. But past success was no guarantee of continued good fortune. The *Lagoda*'s next three outings amounted to two stinging losses and one bare profit.

To help guard against the inevitable losses, ownership, like the lay, was divided into fractions and shared among several owners, sometimes a dozen or more. This arrangement protected capital by scattering it, in the hope that good fortune would balance bad. Jonathan Bourne, one of New Bedford's financial heavyweights, had money enough to own 17 ships. He chose instead to hold shares in 46. Fractional ownership also attracted many small investors, who put their savings into a piece of a whaler and prayed hard that the sea would prove bountiful.

And for some decades it certainly was. Depression, shifting markets, shipwreck, and ordinary bad luck gave whaling its share of hard knocks, but by and large the industry did well. Every year from 1835 to 1860 saw no fewer than 500 Yankee whalers at sea.

Yet, perhaps inevitably, the Yankees declined, as the English had before them and the Dutch and Basques before them. It has become standard to cite increasing competition from petroleum as the reason for the decline, and that competition was indeed partly responsible. But there was more to it than oil.

War, again, played a role. In the early months of the Civil War, the Confederate raiders *Sumter* and *Alabama* caught and burned over a dozen whaling ships in the Atlantic. Ships in the Pacific thought themselves beyond jeopardy, but the war finally caught up with them. In 1865 James Waddell brought the privateer steamer *Shenandoah* into the Pacific. After burning four whalers at Ponape and another in the Sea of Okhotsk, Waddell headed through the Bering Straits into the summer whaling grounds in the Arctic Ocean. In one week in June, Waddell caught 25 whalers, crowding the prisoners onto four of the ships and setting the rest afire. Of course, the war was already over, but Waddell, doubting the veracity of captured California newspapers headlining Lincoln's assassination, kept up his raids until August, when a British ship confirmed the Confederacy's surrender. All in all, the *Shenandoah* put nearly 40 vessels out of the whaling business.

Because the war unsettled business conditions and added new perils to a risky enterprise, many owners transferred their vessels to the merchant trade, sold them to the independent flag of Hawaii, and let the oldest vessels rot at the wharves. And when the federal government offered to buy 40 idle whalers, the owners were happy to sell. As part of a scheme to stop rebel blockade runners, the whalers were filled with stones and scuttled in the mouth of Charleston harbor. The Stone Fleet, as it was called, proved a failure. The sea broke up the wrecks in short order and swept them away.

By the end of the Civil War, the whaling fleet had declined from 514 vessels to 263 and represented only 43 percent of the prewar tonnage. But the industry had recovered from wars before, and this one looked no different. Prices for oil and baleen shot upward and attracted investment. By 1869 the whaling fleet had grown to 336 vessels. But the boom times of old were not to be repeated, in part because of the boom times of the past. Whales were much scarcer. In the late 1840s and the 1850s, many whalers had migrated like the whales, spending the summers in the North Pacific hunting rights and bowheads and coming south into equatorial waters to seek sperms in the winter. The sperm whale, however, was sufficiently depleted that, with sperm oil prices dropping, few ships' masters bothered with the wintertime "between seasons" cruise by the 1870s. The summer hunt was the industry's mainstay, but it too was in decline. Heavy hunting for the lucrative bowhead, with its long baleen and high oil yield, had taken a heavy toll of the species before the Civil War, and the whalers were forced deeper and deeper into the Arctic to find their quarry. The ships penetrated the same leads as did the bowheads, pitting wooden hulls and sail power against the mercurial violence of northern ice.

The reckoning came in 1871. In late August a wind moved the ice against the shore and caught 32 ships near Point Belcher in Alaska. The whalemen waited for the wind to change, but the gale blew steadily and kept the ice in place. The end of the season was fast approaching, and the masters knew what was in store for them. On the way in to the grounds, the fleet had picked up the few survivors of the wreck of the *Japan* who had made it through the winter on blubber and unskinned walrus meat. The masters decided to abandon their ships. All 1,219 people boarded whaleboats and followed a narrow lead 80 miles through the shoreward shoals to Icy Cape, where seven untrapped whalers waited. Although the badly overloaded boats had to cross five miles of open water running a strong storm swell to reach the ships, not one life was lost.

But the property loss proved devastating. Only one vessel of the 32 was salvaged; the rest were sunk, crushed, grounded, or burned by Eskimos. The outright cost of the disaster was $1 million, and insurance rates skyrocketed. More and more owners pulled out. In 1876 the 1871 disaster was repeated. Twelve of 20 ships in the Arctic were trapped by closing ice. Some of the crews chose to run for it; some chose to stay the winter. Not a man of the latter group was seen again.

Yet a remnant of commercial life and Yankee ingenuity lingered. Natural gas and petroleum were replacing whale oil and sperm oil as illuminant and lubricant. Baleen, however, remained in demand. The price per pound rose from $1.02 in 1868 to $2.00 in 1880. The whalemen shifted their home port to San Francisco because of its proximity to the western Arctic. The unending challenge of ice was met in part by adding auxiliary steam engines to the whalers, thus giving them the ability to maneuver independent of the wind. In the thirty years of steam whaling from 1880 to 1910, only 11 whalers were lost, compared to 76 in the previous twelve seasons. Steam also allowed the whalers to push into unknown corners of the Arctic and, finally, to discover the summer feeding grounds of the bowheads in Mackenzie Bay, where the whales could be hunted in open water. Shore stations were also set up along the Alaskan coast, run by white men and crewed by Eskimos who soon became expert with the darting and shoulder guns and went into business for themselves.

This last hurrah of Yankee whaling was monstrously wasteful. The price for whale oil was so poor and that for baleen so rich — up around $5 a pound after the turn of the century — that flensing and trying weren't worth the trouble. The precious baleen was simply cut out and the unflensed carcass set adrift.

The whalers hunted the bowhead and themselves into oblivion. The supply of baleen was so short that industry turned to spring steel as a substitute, and the bottom had fallen out of the baleen market by 1908. The Arctic whaling fleet retired.

It is easy and tempting in telling the tale of the Yankee whalemen to lament in round tones the passing of good ships and strong men. Such nostalgia misses the point. The Yankees did not vanish for a lack of bravery. They vanished for a lack of whales.

Unfortunately, there can be no exact accounting of the toll the Yankees took. The whaleman saw the whale as a resource, a thing far less valuable for itself than for the products to be extracted from it, and whaling records state not the number of animals killed but the barrels of oil and pounds of baleen the ships brought home in their holds. Dividing the total take of oil by the average yield per whale gives a figure, but that figure underestimates the kill. For one thing, not all the oil came to port. Whaling ships at sea bartered oil for provisions and repairs, and traded oil did not enter the import records. For another thing, a goodly number of dead whales never got to the try-pots and thence into the ledger books. Whalemen kept no tally of the carcasses that sank or that had to be abandoned because of heavy weather or of the calves they killed to lure their mothers into harpoon range. These whales died without accounting.

Thus putting a number to the Yankee kill means making an assumption here, estimating an average there, and crossing your fingers everywhere. The result is a rough approximation at best. But it is also a rude awakening. The Yankee whalemen, from their early days as colonists to the boom years and into the arctic twilight, killed about half a million great whales.

In absolute terms, the carnage fell most heavily on the sperm whale, which made up roughly half the total. The other half comprised largely rights and bowheads, with smaller numbers of grays and humpbacks. These latter species, however, were more severely affected than the sperm whale, even though smaller numbers of them were killed. Sperm whales were more numerous to begin with — prewhaling estimates exceed 1,000,000, making the sperm the most numerous of the great whales —

and they existed in many separate populations covering great reaches of ocean. The right whale, bowhead, and gray were far less numerous, and they were concentrated in specific locales where the Yankees, with their methodical ways, could destroy them. And the whales were indeed destroyed.

The first to go was the North Atlantic gray whale, which numbered no more than a few thousand individuals and which was extinct by 1750. The California gray whale survived the Yankees but in much reduced number. The right and the bowhead fared much worse. Before the whalemen went to their work, places like Massachusetts Bay, the Chile coast, and the Northwest Coast harbored tens of thousands of right whales. Today these populations each number only a few dozen. By the time the baleen market fell to spring steel, the bowhead was precariously close to extinction. And although the sperm whale escaped destruction, it received a severe blow.

It wasn't petroleum that ended Yankee whaling. It was the Yankee whaleman himself. He had swept the seas of all that could be killed at a profit with whaleboat, toggle-iron, and bomb-lance. He had drawn the economic rationality of commercial whaling into an irrationality of near-extinction.

•   •   •   •   •

*And nature, too, is bosh — the way
you conceive it. Nature is no temple but
a workshop, and man is the worker
therein.*
— I V A N   T U R G E N E V

The Great Killing

T he story goes that Sven Foyn, sealer by trade and Norwegian by nationality, was moved by the plight of his brother whalemen. Little was doing in the local whaling business. So few right whales, bowheads, and sperms remained in the North Atlantic and adjacent Arctic waters that no one could support himself hunting them. There were other whales in the sea, namely the rorquals, but try as they might, the whalemen rarely succeeded in catching them. Only the slow humpback could occasionally be taken by a whaleboat; as for the blue, finback, and sei, they could readily outdistance the best crew in the quickest craft. Now and then a harpooner might stalk a rorqual and put an iron into it, but the feat often proved more trouble than it was worth. A crew from a Scottish steam-whaler once struck a blue. The animal took out six tubs of line, then pulled three whaleboats and the steamer across the sea for fourteen hours, a Herculean display that ended only when the ship's master decided to bridle the whale down and put his engines in reverse. The whale kept pulling, the line parted, and the animal ran free. Actually the master probably saved himself the further disappointment of losing the whale after killing it. Rorquals commonly sink as they die.

To Foyn this situation was intolerable, an insult to human enterprise and a challenge to the divine enjoinder to rule the earth. The sea offered up an abundance of rorquals held just out of reach. Man simply lacked the tools to pull them in.

Foyn set to thinking. To make up for the speed of a rorqual, the harpooner needed a weapon with far more range than the toggle-iron, and, to break the whale's endurance, he had to inflict mortal or crippling hurt at first strike. Foyn began tinkering with two standard items of whalecraft. One was Greener's harpoon gun, long a favorite of British whalers in the Arctic. The gun, essentially an oversize muzzle-loader mounted on a swivel in the bow of the whaleboat, fired a harpoon and line. The gun injured the whale no more than a hand-cast iron, but it had enough accuracy and range for a practiced boatsteerer to hit from thirty yards away. The other tool was the explosive bomb-lance shot from the shoulder gun, which carried no securing line but had proved useful in finishing off whales already fast. Foyn put the two weapons together, combining the range of the harpoon gun with the punch of the bomb-lance. He fashioned a muzzle-loading cannon, much bigger than Greener's gun, that fired a heavy harpoon tipped with a grenade. As the harpoon crashed into the whale, a glass vial of sulfuric acid broke and detonated the grenade inside the animal's body, shredding its innards with shock and shrapnel. Foyn's cannon could hit accurately at thirty to forty yards, and a strike in lungs or spine would finish off even a blue whale in a matter of minutes.

The drawback to Foyn's cannon was its heft. The weapon weighed close to a ton, far too much hardware for a whaleboat. Foyn mounted the cannon in the bow of a little 30-ton steamer, the *Spes et Fides,* which could make only seven knots, too slow to catch a running rorqual, but steady enough and quiet enough to slip up on a whale and get into position to blast it as it rose to blow. Foyn ran the harpoon line around a winch and through a spring-loaded block called an accumulator, which absorbed the shock of the whale's struggles, much as a fishing rod buffers the pull of a fighting fish, and prevented the whale from getting enough purchase on the line to part it. Once dead, the whale could be winched alongside, where, to keep the carcass from sinking, a hollow tube was thrust into the body cavity and air pumped in to inflate the whale like a balloon.

There was more to Foyn's tinkering than a desire for prosperity. Hard-nosed Christian that he was, Foyn saw nature as so much pagan riot begging for a godly order imposed by man. Rorquals, greatest and swiftest of all the great whales, were wildness supreme. Bringing these quick giants within human grasp extended Christian rule to the heart of nature. Heaven sanctioned the work. While Foyn was testing his cannon, a loop of line snared him and carried him into the sea. Few ever survived such an accident, but Foyn escaped without injury, convinced that God had something in store for him besides an unmarked grave at the bottom of some cold fjord. In thanks Foyn patented his cannon on Christmas Eve.

Foyn took the *Spes et Fides* on its first voyage in 1868 and killed thirty whales. Foyn had accomplished his mission: He had figured out a way to take rorquals and he had shown it could work.

Foyn's technology put whaling on a new course. The cannon killed much more swiftly than the iron and lance, and the catcher did not have to waste time with butchering and processing. Instead, the catcher towed its day's kill to a shore station, took on food and fuel, and went to sea to hunt again. By the mid-1880s, with Foyn's methods imitated far and wide, Norwegian whalers were taking about 1,200 whales each summer, an average of nearly 40 per catcher. The cannon had made the catcher four times more efficient than the Yankee whaler.

The Norwegians replaced the Americans as the world leaders in whaling. They spread out from their home waters in search of hunting grounds and showed just how fast the new technology could destroy the rorquals in range of their shore stations. In 1890 the catch around Iceland amounted to 160 whales; it rose year by year to a high point of 1,257 in 1903, then declined to 54 in 1915, when whaling ceased there for twenty years. The same pattern repeated itself around Spitsbergen, where the take jumped from 57 to 599 in two seasons, then dropped to 58 in only seven more. These declines did not result, as apologists for whaling typically argue, from a drop in oil prices that made whaling unprofitable. When Iceland whaling was abandoned, the price for whale oil on the London market was £5 higher than it had been in the peak year of 1903. No lack of profits spelled an end to the Iceland grounds; it was a lack of whales.

The Norwegians hopscotched around the North Atlantic, from one whaling ground to another. Besides Iceland and Spitsbergen, Norwegian whalemen searched out whales around the Faeroes, the Hebrides, the Shetlands, Scotland, and Ireland. The British, noting Norwegian success, got into the act in Newfoundland, establishing nearly twenty shore stations by 1905 and killing over 4,500 whales in their first ten seasons there. The British, though, continued their habit of relying on outside talent. They hired every Norwegian gunner they could get.

Given the demonstrated killing capacity of the harpoon cannon and steam catcher, the Norwegians would have fast run out of whales unless they found a mother lode of the sort the Yankees uncovered in the Pacific. In the 1890s Norwegian whalemen and sealers, Sven Foyn among them, ventured into the dangerous waters surrounding Antarctica, a region little entered since midcentury. There the explorers came upon rorquals in numbers such as none of them had ever imagined.

Unfortunately, the rorquals of the Antarctic were destroyed before they were counted, and no one knows with certainty how many whales lived there before the whalers went to work. Present estimates, which are statistical extrapolations from whaling data, put the original population of finbacks at 350,000 to 400,000, of blues and seis at 150,000 for each species, and of humpbacks at 100,000. The total came to between 750,000 and 800,000, the greatest concentration of whales in any ocean, a mass of life that dwarfed even the fabled game savannahs of East Africa or the buffalo hordes of the Great Plains.

The vast crowds of Antarctic rorquals were supported by the fortuitous and seasonal intersection of wind, current, and sunlight. In the Southern Ocean, surface currents flow north away from the Antarctic continent until, between 48° and 62° south, they collide with warmer waters from middle latitudes pushed along by the prevailing westerly winds to make the surface current known as the West Wind Drift. This collision of currents, called the Antarctic Convergence, mixes and shifts the northbound waters. Part of the current continues north under the surface, and the rest dives and turns back on itself to return south at a depth beneath 100 fathoms. This subsurface current is warmer than the surface water — 34.5°F compared to 32°F — and it carries abundant nutrient salts taken from the waters of the drift. When the subsurface current strikes the Antarctic landmass, it wells up to the surface, sheds its heat and nutrients, and flows north again to close the circle. This cycle continues all year, but in the springtime, as the ice pack melts and sunlight floods the water, it provides the basis for a great population explosion. The diatom *Fragilariopsis antarctica,* which is a microscopic plant, thrives on the sunlight and nutrients and appears in great abundance. The diatoms become food for a small shrimplike crustacean, *Euphasia superba* or "krill," which clouds the water between 10 and 60 feet deep so thickly that whole patches of sea turn red.

The life cycle of the Antarctic's rorquals is based on the seasonal burst of krill. In the southern spring months of November and December, the whales move south from their wintering areas in the warmer seas farther north. As the polar waters turn red with krill, the whales feed throughout the whole of each long day. Adult blues, as one prodigious example, take in up to four tons of krill a day. The whales have eaten little in the past eight months, and for the four months that the krill last, the whales keep eating, storing the food as fat in their blubber, in the porosities of their bones, finally between muscle strands and in their body cavities against the long winter fast ahead. When April brings autumn storms, the fattened whales move north again, to winter and calve and breed in warmer water, living off the food stowed away in summer. Wintering whales do feed if the opportunity presents itself, but midlatitude oceans produce only a fraction of the plankton found about Antarctica, and opportunity presents itself infrequently.

Sometimes people are heard to wonder why, when the whalers arrived in the Antarctic and began their butchery, the whales didn't simply go somewhere else. The fact is, they had no place else to go. The Antarctic constituted a unique and irreplaceable larder without equal anywhere else. To survive, the whales had to head south, whalers or no.

The Norwegian whalers fairly lusted after the rorquals they found about Antarctica. The whales were numerous and they were huge, running several feet longer than their congeners in the northern oceans. But poor weather made Antarctic whaling difficult, and the cost of ferrying men and equipment 10,000 miles from Norway appeared prohibitive.

Carl Anton Larsen worked out the necessary corporate arrangements. Larsen was making his second voyage into the Antarctic, as ship's captain for a Swedish exploring team, when the vessel went down and the survivors were rescued by an Argentine gunboat. In Buenos Aires Larsen sold executives of an Argentine fishing company on the growth possibilities of Antarctic whaling and talked them into backing a season's trial. In the southern summer of 1904–1905, Larsen brought a catcher and crew from his native Vestfold and steamed south to the natural harbor of Grytviken on South Georgia, where he set up a temporary shore station and killed 195 whales, mostly humpbacks. Larsen's success prompted imitation. More and more catchers steamed south, and shore stations were established on the Falklands, Kerguelen Island, the South Orkneys, and the South Shetlands. During the winter the catchers moved north, whaling the winter grounds from shore stations around southern Africa from Angola to Natal. In the whaling year 1912–1913, 62 catchers worked the Southern Ocean and killed 20,270 whales. Compared to the Yankee industry at its apex, the Norwegians were using only a fraction of the vessels and taking twice the whales. Clearly, whaling had rounded a technological corner.

Of course, there is more to whaling than the tools of slaughter. Whales are not killed for killing's sake but for money's. As the whalemen tried their hand in the Antarctic, industrial chemists were perfecting a process that made the oil of baleen whales a particularly valuable commodity.

At the turn of the century the population of Europe was growing fast, and the demand for such edible fats as butter and lard was expanding apace. Supply fell well short, and the search for alternatives was on. Sperm oil, like the oil from all odontocetes, could not be used. The oil is actually a wax, which the human gut cannot digest and which passes out of the intestines in the same molecular form it enters. The oil of baleen whales, however, is a triglyceride, one of a class of fats that occurs in all plants and animals and that humans can digest readily. Whale oil, though, smells and tastes "fishy" and is a liquid. In 1908, Paul Sabatier and Jean Baptiste Sendérens invented catalytic hydrogenation, a process which adds hydrogen atoms to oil molecules and turns the liquid into a solid. Hydrogenation also made whale oil palatable enough to use as an ingredient in margarine. Twenty years later, chemists of the Margarine Union developed a way of making acceptable margarine from whale oil alone. Thereafter, whale oil became the principal ingredient of about half the lard and margarine eaten in Germany and Great Britain.

In the years before World War I, the slaughter of whales in the service of margarine fell most heavily on the humpback, which made up about 80 percent of the Antarctic catch. The humpback was an easy and well-paying prey. The animal is slow, migrates close to land, and contains considerable oil for its size. In the early days of South Georgia whaling, the catchers simply streamed out of Grytviken harbor, ran down a day's worth of humpbacks, and towed them in to the shore station. The ease of capture and the abundance of whales made for outrageous waste. Humpbacks usually travel in pods, and a catcher could kill several whales if the gunner and crew worked fast. As soon as one whale was dead, it was pumped afloat and set adrift for pickup later. However, with whales so numerous that it was simpler to kill more than to search for carcasses, at least one out of every three or four drift carcasses was lost. The whalers counted only the whales processed, not the whales harpooned, thus understating the true slaughter. The record, for example, gives the kill of Antarctic humpbacks in the years 1910–1911 as 8,294. That is false. The toll actually exceeded 10,500.

Processing was as squanderous as killing. Flensers stripped off only the thick and oily back and belly blubber, leaving the thinner and drier flank, head, and fluke layers on the discarded carcass. The whalers preferred to increase production by killing more instead of making better use of what they had. This strategy made perverted economic sense. It also proved a biological catastrophe.

In the 1909–1910 season, South Georgia yielded 6,197 humpbacks, which amounted apparently to the better part of the local population. The take fell quickly after that. After the 1916–1917 season the kill never again exceeded 500, and by 1926–1927 it had dropped below 100.

The waste offended the British, who had jurisdiction over many of the shore-whaling areas as part of the Falkland Islands Dependencies. They drew up a set of regulations limiting the number of catchers to prevent overkills, prohibiting the killing of cows with calves, and requiring the processing of all blubber. Considering themselves hampered by these rules and disliking the rents and royalities the British exacted, the Norwegian and Argentine whalers escaped English control by resorting to floating factories. These factories were simply ships outfitted with blubber cookers, modern versions of Yankee whalers. The whales were flensed alongside by workmen standing on the carcasses, and the blubber was hauled aboard for mincing and rendering. These factories could not be used in the open sea because the flensers were unable to stand atop the slippery carcasses with even a small swell running. The floating factory had to be anchored in a bay or harbor, where it was less ship than portable shore station. It served its purpose, though; it got the whalers out of British territory and away from British regulation and rent.

During World War I, Antarctic whaling faltered but did not fail, and after the fighting was over, the industry expanded. Britain increased its own whaling efforts and crept up on the Norwegians. But before the English caught them, the Norwegians developed one more important tool of the whaling trade, the last link in the chain first forged in Foyn's cannon.

With shore stations and floating factories, the whalers could exploit only the rorquals one or two days' steaming from harbor. To get at the rest of the whales, the whalers needed a way of working the open sea to the edge of the shelf ice. In 1925 they got it.

The *Lancing,* a floating factory of the Globus Whaling Company of Norway, was fitted with a slipway in its stern through which whales could be hauled·up onto the deck for flensing. The *Lancing* could handle kills even in the open sea. It was the first pelagic factory ship, the beginning of the whaling era in which we live even now.

The prototypical pelagic factory that evolved from the *Lancing* is as ugly as its name implied. The hull is squat, the decks wide, the lines lacking rake or grace. Because of the position of the stern slipway, two smokestacks replace the usual single funnel, one to each side of the ramp. Whales are dragged up the slip in the grasp of giant tongs called the whale claw, or *hval kla.* On the first deck flensers strip the blubber off in long sheets with knives and tackles. Then the naked carcass is winched forward to a second deck area, where workmen called lemmers remove the meat and viscera and break the skeleton with huge steam saws. Intestines go overboard; chemicals in the krill they contain would darken the oil and lower its grade and value. Blubber and bones are minced mechanically and fed into boilers below decks. Meat is sometimes frozen or salted for storage, sometimes rendered. Oil drawn off from the boilers is stored in huge tanks and at intervals during the whaling season transferred to a tanker, from which the factory takes on fuel.

The pelagic factory gave the whalers the wherewithal to move east around Antarctica into the krill-rich waters due south of Africa. There the industry grew almost overnight. In 1919–1920 whale oil production in the Antarctic had amounted to a little less than 275,000 barrels. Incredible oil prices, as high as £93 a ton on the London market in 1920, spurred rapid expansion. By 1931 the number of floating and pelagic factories had increased from 6 to 41, and catchers from 47 to 232. Oil production swelled to just over 3,600,000 barrels.

The Antarctic fleet had expanded about fivefold, but oil production increased almost fifteen times. One might think that the whalers had grown better at their trade, but the actual reasons for the increase have to do largely with how and what the whalers were killing. The Antarctic fleet had actually grown bigger than the simple count of vessels indicates. Ships and catchers were larger, heavier, more powerful. Foyn's *Spes et Fides* displaced only 30 tons; by the early 1920s catchers averaged about 200 tons and sported 700 horsepower. Each year new catchers were added, ever bigger and ever faster, able to cover more water in a day and kill that many more whales. Thus, while the number of catchers increased by five between 1919 and 1931, the kill went up by eight, from 5,441 to 40,201. The factory ships also grew, nearly doubling in the 1920s to more than 10,000 tons to accommodate the larger kills brought in by the bigger catchers.

But the pronounced increase in whale oil production came also from the killing of bigger whales. The pelagic factory ship allowed operation at the edge of the shelf ice, which was the principal feeding zone for the blue whale. By the early 1930s the blue whale made up three-quarters of the Antarctic kill, having replaced the much-depleted humpback as the prime object of slaughter. There was good economic reason for the whalers' interest in the blue. It took just as much fuel and just as many ship hours to hunt down a blue as a humpback, but the blue yielded much more for the expenditure. The typical humpback weighed a little over 30 tons, while an average blue hit 80 to 85 tons, with the really big individuals of close to 100 feet in length running up toward 175 tons. Each dead blue was a bonanza of raw material, and the whalers pursued blues with a will. By the 1930–1931 season, over 125,000 blues had been killed to supply European kitchens with margarine and lard.

In no way could the blue whale support such slaughter. The toll already equaled nearly the original population. Of course, whales were being born all the while, but the rate of reproduction — one calf for each breeding-age female every two or three years — was but a fraction of the pace of killing. The blue, like the humpback before it, was living out a prophecy made some twenty years before by R. Collett in the *Norges Pattedyr:* "Foyn's harpoon gun still pursues its march of death through all the oceans, and with unprecedented force. For some time yet great wealth may be got from the sea by its means; but the heavy toll levied upon a group with so comparatively limited a power of reproduction as the whales will inevitably reduce the stocks."

Fortunately, both Norway and Britain, the leading Antarctic whaling nations, started to pay attention to the havoc they were wreaking. The English were concerned that the exploitation of Antarctic resources, one of which was whales, was proceeding too rapidly and that bust would soon follow boom. Little was known of the region in general, and of whales in particular ignorance was almost complete. Britain formed the Discovery Committee, which, among its other tasks, conducted the first systematic investigations into whale migrations. As for the Norwegians, they expanded the statistical reporting begun by Professor Sigurd Risting, who even before he became secretary of the Association of Whaling Companies had been gathering and publishing whaling data in a fisherman's gazette. The Norwegians collected information on the size and number and species of killed whales, sex distributions, number of pregnant females, size of fetus, and so forth, the first such collection of cetological data in history.

The Norwegian data showed that the average size of killed blue whales was falling each year, a clear indication that whaling had cut deeply into the blue whale population. The whalers, though, slowed their pace not a bit. They responded to no scientific handwriting on the wall, but to the sound of money in the cash register. Little did they know that the fickle world of finance had a comeuppance in store for them.

The whalers killed so many whales they nearly did themselves in. Increasing oil production glutted the market, dropping the price just as the Depression took a crushing grip on Europe's houses of financial cards and toppled one after another. Many whaling companies failed. Business prospects were so bad that few firms even bothered with the 1931–1932 season.

The Norwegian and English whalers did what any collection of properly

Blue whale (*Balaenoptera musculus*), flank view (top); dorsal view (bottom)

self-interested capitalists would: they divided the territory. Company representatives agreed that the only way out of their predicament was to limit oil production and to apportion that limit among themselves. For the first time, opening dates were set for the Antarctic season: October 10 at South Georgia and October 20 elsewhere. The production total was parceled out to the individual companies in terms of the number of barrels each could produce. The oil quota in turn determined how many whales that company was allowed to kill. One blue whale was expected to yield 110 barrels, as would three humpbacks or five seis. If a company would extract more than 110 barrels from each blue whale or its equivalent in other rorquals, its production quota was increased by the amount of extra production — a provision won by the companies that prided themselves on their efficiency.

The agreement held for the next two seasons. Although the kills were only about 25,000 whales and oil production fell, the price of whale oil continued to drop until it hit £12, its lowest point yet. Then business took a turn for the better. The price edged up to £20 and the agreement started to come apart. While it remained in the interests of all whalers to maintain the combine and to keep production down and prices up, it was in the self-interest of each to take all the oil he could and sell it while the price stayed high. Thus, 21 companies entered into agreement for the 1935–1936 season, but eight others remained outside the fold. This apparent return to the chaos of the 1920s bothered the government of Norway, which assigned quotas to its companies. For the 1936–1937 season the Norwegian and British governments together agreed on quotas for their companies and set new dates for the Antarctic season: December 8 to March 7. In addition, the number of catcher boats in each pelagic expedition was limited according to the tonnage of the factory vessel.

The production limits, though, had little effect on the progressive decimation of the Antarctic rorquals. Kills went up even as the stocks declined. Yet these agreements did offer some small reason for hope. They were indeed limits, even if more in theory than in fact, and limits were something that whalers had never before adopted.

Curiously, the nation that laid the legal bases for the control of modern whaling was the very country that had given the world the tools for slaughtering rorquals: Norway. In the early days, when Norwegian whalers were still working their home fjords, complaints from fishermen that the ruckus and slaughter were scaring away the fish and public concern that the whales might be exterminated led the Norwegian government to close its north coast to all whaling in 1904, thereby creating the first whale sanctuary. In 1929 the Norwegian legislature passed the Act on the Taking of Baleen Whales, which, given the extractive temper of the times, was an extraordinary document. The law gave the government power to control practically all the activities of Norwegian whaling fleets on the high seas. It prohibited the taking of right whales, of calves, and of females with calves. Minimum-size limits were set for each species, and factory ships had to keep daily records of kills. To enforce these regulations, every factory vessel carried a government-appointed inspector who answered only to the Minister of Commerce. In the 1930s a second inspector was added, to ensure that no dirty deeds were done while sleep closed watchful eyes.

International regulation of whaling was slower in coming. Soon after the development of the stern-slip pelagic factory freed the whalers from the shore and gave them the run of the high seas, the League of Nations decided that regulation by all concerned nations was called for. No convention was drafted until 1931, however, and four more years elapsed before enough signatures were appended for it to enter force. The convention borrowed heavily from the Norwegian whaling act. It too prohibited the killing of right whales, calves, and females with young, and it required that industrial and biological statistics be sent for tabulation to the Committee for Whaling Statistics in Oslo. The convention was notably shy on enforcement. Each signatory government was to police the agreement in its own fleet, but the convention did not call for inspectors nor did it dictate penalties for violations. The sole requirement made by the convention was that all whaling vessels flying the flag of a signatory government be licensed by that government, a requirement that did nothing more than keep a few extra bureaucrats busy issuing papers and collecting fees.

The convention did not even slow the international rush to plunder the Antarctic. Panama and South Africa began pelagic whaling, as did the United States, which operated a pair of factory ships in the Antarctic and off Australia. Two other nations made a bigger and more destructive entry: Japan and Germany. Hitler, in preparation for spreading the glory of the Reich from the Pyrenees to the Urals, ordered the stockpiling of strategic goods, among them edible fats. Germany sent out its first pelagic fleet in the 1936–1937 season and hired two Norwegian expeditions as well. Japan, too, was planning for war. The Japanese had picked up Norwegian techniques by copying Russian whalers in the Sea of Japan and had established some forty shore stations in the eastern Pacific. In 1934 the Nippon Suisan Company bought a used factory ship from the Norwegians and sent it to the Antarctic. Within a few years Japan had six factory ships, which worked the North Pacific and the Bering Sea between Antarctic seasons.

Prompted largely by Norway, the whaling nations met in London in June 1937 to strengthen the 1931 convention. Japan, which refused to abide by any regulations, did not attend. This meeting drew up the International Agreement for the Regulation of Whaling, which added significant new rules. Protection was extended to gray whales as well as rights. Minimum-size limits were set for the first time in international law: 70 feet for blue whales, 55 for finbacks, 35 for humpbacks and sperms. All pelagic whaling was prohibited between the equator and 40° south, establishing the latter line as the boundary of the Antarctic. The agreement also incorporated several provisions aimed at reducing the total kill by eliminating waste. It required the full use of the whale carcass; oil was to be extracted from meat and bones as well as blubber, thus raising production without raising the kill. It also demanded that catchers kill no more whales than the factory they served could process within 36 hours, a rule designed to prevent waste due to spoilage. Each factory ship was to carry inspectors, but enforcement remained in the hands of the individual nations. The Norwegians and the British tended to be scrupulous. The South Africans, by contrast, violated the agreement almost as soon as they signed it, by beginning whaling in the Antarctic a full month before the opening date of December 8.

The agreement, for all its seeming improvement, had no effect on the kill of whales. The 1937–1938 season witnessed the killing of the most great whales to date: 54,835 worldwide, of which 46,039 died in the Antarctic. Yet, even though the 1937–1938 kill exceeded the previous high point of 1930–1931 by more than 11,000 animals, oil production actually fell 60,000 barrels below the 1930–1931 figure. Blue whales were growing scarcer fast. In a period of less than ten years the catch had dropped by half, even though bigger and more powerful catchers were hunting the same waters. Finbacks had replaced blues as the species killed in the greatest number.

The decimation of the blue whale alarmed the International Council for the Exploration of the Seas, at whose behest the whalers again met in 1938, again without Japan. The members of the council wanted the take of blue whales cut. Nazi Germany refused even to consider such a limit. The Norwegians, anomalously the leading whalers and the leading conservationists, proposed that the number of catchers be reduced, but this proposal also made no headway. Finally two amendments to the 1937 agreement did get past the Germans. Protection was extended to the humpback in Antarctica; of course, the species was already so scarce that it was only a commercial sideline. Also, a portion of the Southern Ocean was closed to hunting to create a whale sanctuary.

In 1939 Germany found Poland more interesting than the Antarctic and sent out no whaling expeditions. The cold southern waters were left to the Japanese, the Norwegians, the British, the Argentines, the Panamanians, and the Americans. Numbered among the whales killed in this final year of the first round of Antarctic whaling were nearly 900 humpbacks. So much for protection.

World War II put an end to pelagic whaling in the Southern Hemisphere. Naval war exacted a fierce toll on whaling vessels, particularly factory ships, which offered big and slow targets to submariners and pilots of the Axis and the Allies alike. In one of the marvelously ironic twists of human history, our war gave the whales peace.

In the 1930s the two forces that shaped the present struggle over whaling came to the fore: the destructive force of whaling and the protective force of conservation and regulation. The precedents in the Norwegian whaling law and in the 1937 agreement have proved basic to the effort to save great whales that goes on even today. Yet, in the face of the various conventions, protection, agreements, and regulations, the industrial pace of whaling accelerated through the 1930s, just as it had through the 1920s.

At the time the war broke out, factory vessels averaged almost 14,000 tons, nearly three times larger than the first such ships. Catchers were bigger, too, running to 300 tons and carrying in excess of 1,000 horsepower generated, after 1937, by diesel instead of steam. The diesel was more fuel-efficient than steam and much faster. The diesel's clatter alarmed whales and sent them running, but the new catchers were swift enough to give chase and to overtake the whales as they tired. Gunners found that the chase made for better shooting than the old way of quiet stalking, since a whale in a sprint brings more of its back out of the water than it does while cruising.

The toll taken by the bigger catchers and larger factory ships was immense. According to Norway's Committee for Whaling Statistics, the worldwide kill between 1900 and 1940 was 794,000 whales, of which 510,000 came from the Antarctic. Remember: these figures refer only to processed, reported kills. At least one whale out of every five harpooned was lost to bad seas, sinking, or spoilage. Add this 20 percent loss to the Norwegian data, and the kill rises to 977,000 worldwide and 612,000 in the Antarctic. The numbers mean that modern whalers had killed twice as many whales as the Yankees in about one-third the time.

Historically, whaling represents a series of near-extinctions, from species to species and ocean area to ocean area, each destruction proceeding faster than the one before it. It took about 400 years to reduce the black rights to a few thousand survivors. The sperm whales of the Atlantic absorbed about a century of abuse; those of the Pacific, somewhat less. The bowheads of the western Arctic were hunted into near-nothingness in only fifty years. All this destruction occurred without the cannon and the steam catcher. With them, the whalers took most of Antarctica's humpbacks in two decades and shifted their attention to the blue whale. In only fifteen years catches of blue whales declined to half what they once had been. Each season was a quantum jump toward commercial extinction.

The original abundance of the Antarctic, the massive herds of rorquals that turned the horizon into hoarfrost with their spouts, must have overwhelmed anyone who beheld it, much as the American pioneers could not really believe that there was an end to the bison clogging the prairie. Perhaps because of this original abundance, whaler and scientist alike overestimated the productivity of the animals. Surely, they assumed, organisms so numerous must produce young in great numbers. They assumed wrong. In fact, rorquals reproduce only very slowly. Sexual maturity occurs around six years of age in cows, and thereafter a female can produce one calf every two years. Often, though, the cows skip a year or two in the cycle, bearing their single young at a three- or four-year interval. No one is yet certain how many calves a cow can bear, because rorqual life span remains unclear — somewhere between 25 and 50 years — and because it is unknown whether the cows go through menopause and become infertile before death. In any case, many contemporary cetologists figure that killing more than 2 to 6 percent of a rorqual population will cause a slow but inevitable decline. The early whalers took whales at up to ten times this pace.

Admittedly, the whalers of the 1930s were not graced with the present partial knowledge of cetacean population dynamics, but they could have seen what they were doing if they had simply opened their eyes. The statistics of the industry told the tale. In the early 1940s, Johan Ruud, a highly respected cetologist, sat down with the whaling data to do a bit of calculating. During the previous ten years, annual catches had gone up, except for the short-lived setback occasioned by the Depression, but so had the number, speed, and efficiency of the catchers. Were the whales as abundant as ever or were the catchers killing more of them? To answer the question, Ruud calculated the average number of blue whales or their equivalent in other rorquals caught by one catcher hunting for one day. Ruud found that the average rose from 1929 to 1934, then by 1938 fell below the 1929–1930 figure even though the catchers grew larger and more efficient all the while. Ruud concluded that "the stock of whales in the Antarctic has decreased below a level which can be balanced by increased [whaling] activities." The whalers had to work harder for each whale they killed because the whales were fewer. In economic terms, the whalers were earning diminishing returns. In folkloric images, they were killing the goose that laid the golden egg.

It is mistaken to assume that if the whalers had realized the situation fully, they would have changed their obviously irrational ways. A cliché of liberal ideology has it that an entrepreneur profiting from a resource will strive to protect that resource in order to make profits over the long run. Thus, the argument continues, market forces contribute to the preservation of the environment. This may be good corporate public relations, but it is poor economics. An entrepreneur's interest is to increase his own wealth, and he rarely thinks ahead more than five or ten years. The whaling capitalist cares not a fig whether there will be whales in the sea in fifty or a hundred years. His concern is solely whether the whales will last long enough for him to pay off the debt on his factory ship. He will keep on killing as long as some gain can be had, then will either scrap the vessels or shift them to more lucrative employment. Market forces provide no check to extinction; they actually contribute to it. Greed preserves only itself.

But greed is hardly the exclusive property of capitalism. As the whaling events of the postwar era unfolded, the nominally socialist Russians proved themselves as rapacious as the wolves of Wall Street.

While the battles of the Second World War blazed, plans were laid to rebuild the whaling industry when peace made the seas safe for commerce again. Following talks between the British government and the Norwegian government-in-exile, all the countries signatory to the 1937 agreement, with the obvious exception of Germany, met to discuss improved regulation of whaling. All agreed that the prewar take of whales had been excessive and that the kill must be reduced. Their resolution was, of course, academic since the war prevented all whaling.

But when the war did end, the time had come to establish a mechanism of control. The whaling nations met in Washington, D.C., in late 1946, and, after listening to a number of alternative proposals, adopted an arrangement put forward by the United States. That document, the 1946 International Convention for the Regulation of Whaling, established the International Whaling Commission (IWC). The original members of the commission were Argentina, Australia, Brazil, Canada, Denmark, France, Holland, Iceland, Mexico, Norway, Panama, South Africa, the Soviet Union, the United Kingdom, and the United States. Japan adhered to the regulations under the watchful eyes of American inspectors and, after regaining sovereignty, joined the IWC in 1951. New Zealand also joined later.

The preamble to the 1946 convention stated why the IWC was being established. The contracting governments recognized that it was in their several national interests to safeguard the "great natural resources represented by the whale stocks," and they likewise recognized that their interests had been ill served by the "history of whaling [which] has seen the overfishing of one area after another and of one species of whale after another to such a degree that it is essential to protect all species of whales from further overfishing." Destruction could be prevented and preservation enhanced by wise management based on taking no more whales than the number provided by natural increase and by hunting only the species able to sustain continued slaughter while granting the depleted varieties time to recover their former abundance. However, the preamble cautioned, these worthy purposes were to be met "without causing widespread economic and nutritional distress" and without diverting "the orderly development of the whaling industry." In short, the International Whaling Commission was committed to serving both God and Mammon.

The guiding principle behind the 1946 convention was to establish a permanent regulatory body which suffered from none of the uncertainty of the annual agreements of the 1930s but which enjoyed an equal measure of flexibility. The convention thus comprised two portions. The first laid out what the commission was to do and how it was to do it. This section could not be amended. The second could be. Called the Schedule, it detailed specific whaling rules and regulations, which IWC members could alter as times changed and the need arose.

The actual working members of the IWC were the commissioners, one from each nation, aided by any experts or advisers they cared to bring with them. The commissioners met at least once a year to decide the matters before them by simple majority, except for changes in the Schedule, which required three-quarters.

Permanent though it was, the IWC came close to being as weak as an international organization can be and remain an organization. Staff consisted of a single secretary and assorted clerical help. The IWC had no funding of its own and unless granted special allocations by the members was incapable of supporting the research it was empowered to sponsor. As a result, the data needed for decisions about quotas, protections, and regulations had to come from industrial sources, which were not exactly disinterested. Minimum-size limits and species protections were compromised by the members' right to grant permits for the killing of any cetacean for "purposes of scientific research subject to such conditions as the Contracting Government thinks fit." The commission was specifically enjoined against limiting the number or nationality of factory ships or dividing the quota among the various fleets. Finally, any nation displeased with the quota or restriction could escape the stricture by filing an objection to it within ninety days of adoption. An objection kept the regulation from entering into force for another ninety days, in which period the other members could also object. Thereafter the regulation bound only those who had not objected to it. In effect the objection procedure granted each nation a veto over the IWC's doings.

The original Schedule derived from the conventions and agreements of the 1930s. Government inspectors watched over both factory ships and shore stations. Calves and females with calves were not to be taken. Gray whales and all right-whale species were protected worldwide, except when killed by or for aboriginal hunters, and humpbacks were protected against pelagic whaling in the Antarctic. An aggregate quota for all rorqual species in the Antarctic was set in blue whale units, or BWU, with one BWU equaling 1 blue whale, 2 finbacks, 2½ humpbacks, or 6 seis — an equation derived from supposed oil yield. The original quota was 16,000 BWU, which the whalers saw as a considerable cut from the almost 30,000 taken in the 1938–1939 season. Opening and closing dates for the Antarctic season were set at December 15 and April 1 inclusive, with the season to end sooner if 16,000 BWU were killed beforehand. The Antarctic whale sanctuary was retained. Pelagic operations for baleen whales were prohibited in part of the Arctic, in the Atlantic and Indian oceans north of the 40° south Antarctic boundary, and in the western half of the South Pacific. Minimum-size limits for pelagic whaling were set at 70 feet for blues, 55 for finbacks, 40 for seis, and 35 for humpbacks, with lower limits allowed for shore-station killing for human consumption. No factory ship or shore station could operate more than six months in a row in one area. Carcasses could remain in the water no more than 33 hours before processing, and the whole carcass except for viscera and sperm whale meat had to be processed unless the whale was destined to feed humans.

Certainly the men who drafted the original Schedule were at least somewhat naive, unaware that the survival of the large rorquals would require controls far stronger than they were even asked to consider. Yet it is impossible not to detect the distinct scent of purposeful duplicity in these rules. Again and again the regulations are form without substance, legal creations that give the appearance of conservation while in fact conserving nothing.

Not one of the areas closed to pelagic whaling contained enough rorquals to support high-cost pelagic whaling. Closing them cost the whalers nothing. In the waters north of the Antarctic boundary, for example, the rorquals were scattered and so thinned down by the winter fast that they yielded little oil for all the trouble it took to find them. The six-month limitation on whaling operations meant almost nothing, since migrant species rarely spend that much time in one vicinity anyway. The quota applied only to baleen whales in the Antarctic. Whales of other species in other areas could be taken in any number.

The most egregious hypocrisy of all was the minimum-size limits. These limits supposedly served the same purpose as keeper sizes for game fish: to allow the animals to reach sexual maturity and reproduce themselves before they were killed. The size limits on whales failed absolutely at their purported task. Had the framers of the 1946 Schedule wished to set a rational size limit for the blue whale, they could have done so simply by consulting Table α in *International Whaling Statistics* No. 16 (1942), which shows the lengths of over 15,000 pregnant blue whale cows killed in the Antarctic in fifteen seasons before the war. Few of the cows were under 78 feet, with by far the greatest number in the range of 83 to 87 feet. This trend was corroborated by a Norwegian study published in 1948, which showed the average length at sexual maturity for female blue whales to be 78 feet. To allow the whales time to breed, carry the fetus for ten months, and suckle the calf for seven more before weaning, the size limit should have been no shorter than 83 feet and preferably 85 feet. Instead, it remained at 70 feet, with the result that blue whales were killed even before they bred, much less reproduced. The 70-foot limit was a mockery.

Nor was the blue whale alone. The same mockery was applied to the finback. Of almost 16,000 pregnant cows tabulated, exactly 12 — or 0.08 percent — were under 60 feet. Obviously the 55-foot limit allowed finbacks to be killed before they were anywhere near maturity. The limit should have been at least 70 feet.

Greed prevented the setting of effective limits. A finback limit of 70 feet and a blue whale limit of 85 would have cut the kills of those species by 50 percent and 75 percent respectively. The whalers never would have agreed to such reductions.

The 70-foot and 55-foot limits gave the appearance of conservation but caused the industry the least possible inconvenience. Gunners disliked all limits. In the heat of the chase with a 25-knot wind in his teeth and a strong swell running, even the most experienced gunner couldn't tell an 80-foot blue from an 85-footer. The gunners, though, through a quirk of whale behavior, could live with the 70- and 55-foot limits better than any others. Juvenile rorquals, like many young animals, are playful and inquisitive, often approaching boats and frolicking about them. Only as they age do they become standoffish and easily spooked. In blues the change from inquisitiveness to fear occurs at about 70 feet; in finbacks, about 55. A gunner on whales who spared the curious and killed the fleeing only occasionally shot undersize animals. Thus, the size limits served the convenience of the whaling industry, not the biological realities of the whales.

The return to the Antarctic provided a quick lesson in those realities. The demand for fats in Europe had driven the price of whale oil up over £100, and in the 1946–1947 season, 20 pelagic factories, 40 shore stations, and some 250 catchers went to work. The whalers had been looking forward to this season with great expectations. The Antarctic, they were sure, would be simply brimming with rorquals. After all, there had been no full-scale whaling effort for five seasons. A statistical study by the cetologists N. A. Mackintosh and J. E. F. Wheeler showed that blue whales reached maturity at two years of age. With such a short generational time and with such a long pause in whaling, the blues should have recovered significantly from the overzealous hunting of the 1930s. But the 1946–1947 season proved a disappointment. The kill of blues amounted to 9,302, a considerable number of whales, to be sure, but well shy of the 14,081 taken in the 1938–1939 season. The decline was not simply a matter of bad luck or worse weather. In 1947–1948 the Antarctic fleets gained fifty catchers, but the kill of blues fell further, to 7,157.

The declining catches could mean that the blue whale reached sexual maturity at an age greater than two. Johan Ruud explored this possibility by correlating growth ridges in the baleen plates of a sample of blue whales with their sexual maturity or immaturity, while P. Ottestad, a colleague of Ruud, refined Mackintosh's and Wheeler's statistical study. Ruud's and Ottestad's findings agreed: blue whales usually matured at five, but sometimes not before seven. They published their results in 1950, and two years later a confirming study came from cetologists in Japan.

This research heightened the meaning of the whaling data. The blue whale had been hunted into a severe decline from which its delayed maturity and low rate of reproduction could rescue it only very slowly. If the IWC whalers were in truth committed to avoiding another repetition of whaling's oft-repeated overkill — as they claimed in the preamble to the 1946 convention — then hunting of the blue whale had to be curtailed.

Instead, whaling intensified. England subsidized the rebuilding of its fleets with loans, grants, and price supports, as did the Netherlands, which reentered pelagic whaling after an absence dating from the late eighteenth century. The war had cost Japan all six of its factory ships, but they were replaced with modern vessels at the urging of General Douglas MacArthur, who wanted Japan to start feeding itself again. Russia, too, was getting into pelagic whaling as part of Stalin's build-up of Soviet naval and maritime capacity. And Norway, pioneer of the Antarctic grounds, remained very much in the picture.

Given the increasing competition for the Antarctic's remaining whales, the whalers were not about to listen to anything as patently silly as reducing the catch of whales. Indeed, one of the first actions of the IWC, taken at its initial meeting in 1949, was to remove protection from the humpbacks in Antarctica and allow a quota of 1,250 a year. This species quota proved not to the whalers' liking and was dropped in favor of a four-day humpback season, which, despite steadily declining catches of the obviously imperiled species, remained open until 1963. As for the blue whale, the slaughter continued as the take fell from its postwar high of 7,781 in 1948–1949 to fewer than 1,500 ten years later. The finback became the mainstay of Antarctic whaling, dying at a rate that soon exceeded 25,000 a year.

There was economic reason to this environmental madness. Extremely high prices for whale oil at war's end had drawn considerable investment to whaling and tied up that investment in large capital outlays. According to George Small, who has studied the economics of whaling in detail, a pelagic fleet comprising a factory vessel with a dozen catchers cost about $12 million in the early 1950s and twice that by the end of the decade. A whaling fleet could earn back its investment only by killing whales, and enough had to be killed each year to pay off the huge operating costs of getting to the Antarctic and back and to retire a portion of the debt. Once the whalers sank their money into whaling, they were caught in the trap of their own indebtedness. They cared nothing about the whales. What concerned them was the investment bankers and stockholders breathing down the back of their corporate neck.

The way whaling was conducted in the 1950s added to the madness. On opening day all the catchers of all the fleets started killing and kept killing as fast as they could. At week's end each fleet radioed its kill to the Committee for Whaling Statistics in Norway, which kept a running tally. When the take neared the quota, the committee declared the day on which it would be reached and the whaling season closed. Thus all the fleets raced against each other for a share of the available whales. As long as the weather held, catchers and factories went at it full-bore round the clock, a grisly and exhausting ordeal the crewmen called the Whaling Olympic. Declining whale oil prices in the mid-1950s intensified the competition, for each fleet had to try that much harder to kill that much more of the quota just to stay even.

Whaling was heavily overcapitalized. The whalers, though, did not bow out. Like poker players with poor hands and too much in the pot, they kept raising the stakes in hope of making back what they had already put in. Investment increased. All the whalers kept buying more and fancier gear to put them one up on all the other whalers. Catchers doubled in size and horsepower, averaging over 700 tons and 3,000 horsepower. They carried a 90 mm cannon firing a 150- to 200-pound harpoon tipped with a new flat head designed by the Japanese to carry down into the whale's body even on a glancing blow. The Japanese also perfected the use of sonar to track whales beneath the surface and developed ultrasonic gear for herding whales into cannon range. Given these tools, the catchers could kill almost every whale they could find. Catchers did nothing but kill and mark the carcasses with radio transmitters and radar reflectors. Towing was left to buoy boats, which ferried the carcasses to the factory ship. The Japanese liked technology. They equipped a scout boat with a bathythermograph for measuring the temperature of ocean layers to determine where krill blooms, and thus whales, would appear. The Russians liked bigness. Abuilding in their shipyards was the *Sovietskaya Ukraina,* biggest pelagic factory ever, a ship of 715 feet in length, 85 feet in beam, and 46,000 tons in summer deadweight.

The blue whale in particular was caught in a crunch created by the economics of whaling, the BWU quota, declining Antarctic stocks, and the Whaling Olympic. Blues, finbacks, seis, and humpbacks are distinct creatures, each as different from the others as an apple is from a pear, but the BWU, in defiance of all the logic of arithmetic and biology, tossed them in one hopper and added them up. Blue whales were so large and therefore so valuable that the whalers wanted them badly, and the fewer whales there were, the more the whalers wanted them. It took the same amount of money, time, and effort to kill six seis as six blues, but the blues yielded, according to the BWU standard, five times more raw material. No gunner in his right economic mind, racing the clock and the several hundred other gunners in the Antarctic, would pass up an opportunity to kill a blue, even if he happened to be blasting the last one on earth.

Meeting a few days in advance of the IWC's session in June 1953, the various members of the Scientific Committee, which was responsible for analyzing statistics on populations and ensuring that quotas and rules made good biological sense, agreed among themselves that the declining catches of blues indicated trouble. The committee wanted to delay the opening date for hunting blues to January 15 and to close the Atlantic area of the Antarctic to blue whaling. The Dutch delegation threatened objection to any closure, and the proposal was dropped. The season opening was delayed to December 15, a thoroughly meaningless compromise. The whales were simply given two weeks extra to fatten before they were killed.

At subsequent meetings the Dutch continued to oppose any limitation on blue whaling, arguing that the Norwegians erred in their pessimism, that the species required no protection, that thousands upon thousands of blues remained. The loudest and most authoritative spokesman for the Dutch was E. J. Slijper, one of the most prominent cetologists in the world. Slijper never changed his curious tune. He wrote later, "All the arguments [for reducing whale catches] are still based on very little evidence, and . . . whalers can and will restrict their activities only on the most incontrovertible arguments." It is hard to imagine an argument more incontrovertible than the extinction to which blind stubbornness like Slijper's was fast consigning the blue whale.

Holland, though, was not alone in its destructiveness. At the 1955 IWC meeting the commissioners cut the quota to 15,000 BWU, a symbolic gesture, then opened the Antarctic whale sanctuary to hunting for the first time in almost twenty years. The results of the opening were quizzical. The total kill increased by 1,000 animals to 38,580, but the take of blues dropped 500 to 1,614. This proved that the argument that the sanctuary harbored more than enough blues to provide the species a buffer against extinction was hogwash. To protect the surviving whales, the Scientific Committee voted that the quota be cut to 11,000 BWU. The Netherlands again proved intransigent, indicating that any quota below 14,500 would draw an objection and lead to a free-for-all. The Dutch blackmail worked. The quota was set at 14,500 BWU for the 1956–1957 and 1957–1958 seasons.

The Dutch seemed insistent on playing out the role of bad guy to its last vengeful absurdity. In June 1958 they argued for and won a return to the 15,000 BWU quota. They also torpedoed a plan to end the Whaling Olympic. Since the IWC could not divide the quota officially, the pelagic-whaling nations met privately to allocate the quota among themselves. The Netherlands demanded a share twice as large as its average catch and thereby ruined any chance of a gentlemen's agreement.

At the 1959 meeting the IWC very nearly blew apart. The Norwegians felt something had to be done and now. The blue whale was disappearing, and already catches of finbacks had begun the inevitable nosedive. The Norwegians voluntarily agreed to an even lower proportion of the unofficial division of the quota in order to accommodate the Dutch demand for a bigger share, but the Netherlands delegation spurned the generosity, walked out of the meeting, withdrew its IWC membership, and announced its intent to begin whaling in the Antarctic at its leisure. As a result, the IWC set no quota and opened the whaling season ten days early to keep the Dutch from getting too much of a head start on everybody else. Norway, properly chagrined, also withdrew from the IWC, but stated that its fleets would adhere to all regulations, a move whose moral force kept the next Antarctic summer from becoming more of a chaos of whale killing than it turned out to be.

Norway returned to the IWC in 1960. At that same meeting, the Scientific Committee, largely at the prompting of U.S. Commissioner Remington Kellogg and the committee's chairman, Johan Ruud, admitted that it could not estimate the whale populations of the Antarctic accurately and come up with sound quotas. A remedy was proposed: the hiring of three expert biostatisticians, drawn from nations uninvolved in pelagic whaling to ensure impartiality, who were to analyze the available data and to determine the kill each species could sustain without threat to its existence. The scientists who made up the so-called Committee of Three were Sidney Holt of the UN's Food and Agriculture Organization, K. Radway Allen of New Zealand, and Douglas G. Chapman of the United States. In that same year, the Netherlands rejoined the IWC, the pelagic whalers negotiated an acceptable division of the quota, and the greatest whale kill in history was concluded: 66,097.

The Committee of Three reported to the IWC at the June 1963 meeting. The assessment was grim. Antarctic whale stocks had fallen off dramatically, and the very existence of the blue whale and the humpback was in jeopardy. The committee recommended that the blue and the humpback be protected and that the quota be set no higher than 5,000 BWU. The commissioners did vote protection for the humpback; catches of that species were too insignificant to worry about their loss. But at the rest of the committee's recommendations the assembled whalers balked. They wanted the quota reduced only very slowly. Thus they granted protection to the blue whale except in one area, but that exception gutted the protection, for the open zone had been supplying the majority of captured blues in recent seasons. The Japanese, who proposed the exception, came up with what they claimed was a scientific explanation. This zone contained not blue whales, but another species, the pygmy blue whale. The Japanese argued that this species had to be managed separately from the regular blue whale and that closure of the pygmy's range would violate the spirit of the IWC charter. The commission acceded to Japanese wishes, less because of their substance — the pygmy blue was a cynical fiction — than because of the threat of a veto in the event of refusal. The zone was left open to the hunting of blue whales and the quota was set at 10,000 BWU.

Oddly enough, this last hurrah of Antarctic whaling actually enhanced the Committee of Three's status. Allen, Holt, and Chapman predicted a pelagic take of about 8,500 BWU for 1963–1964, well below the set quota. The actual catch was 8,429. This close call, Chapman says, impressed the IWC commissioners. Still, it was not enough for them to cut the quota below 5,000 BWU as the committee wanted. The Soviet Union argued vehemently against a quota below 10,000 BWU, a stand which, in view of the previous season's shortfall, meant that the Russians wanted whaling to continue unrestrained. Japan and Norway lined up behind the Russians, content to let their communist counterparts carry the ball for them. A motion passed, over the sole negative notes of Japan and the Soviet Union, to protect the blue whale throughout the Antarctic. But no quota could be agreed on, and Japan vetoed protection of the blue whale by filing an objection.

The Antarctic season proved poor. The whalers, unrestrained by a quota, took 32,500 whales, but most of them were seis, with the result that the pelagic catch in BWUs fell to under 7,000. Hoping to pick up the pieces of an apparently broken IWC, the United States called a special meeting in London in May 1965. The Japanese arrived with the expectation that the Soviet Union would continue its intransigence. But, in the intervening months, a typically Bolshevik shakeup had rattled the Ministry of Fisheries and the new delegation came to the meeting ready to conform to the inevitable. The Japanese were so taken by surprise that they bolted the meeting to cable home for instructions. When the dust settled, the Japanese, too, capitulated. The quota was cut to 4,500 BWU, and the blue whale was given protection throughout the Antarctic. The move was a bit belated. The previous season's catch of blues came to 20 — of which, in an ironic twist, not one fell to the catchers of the Japanese, whose objection had allowed the killing to go on one more year. The species was commercially extinct in the Antarctic.

Protection of the Antarctic blue whale, symbol that it was, marked the end of one phase of modern whaling and the beginning of another. Scene and actors changed again as they had many times before.

The Antarctic heyday was well past. The finback was fast approaching the commercial extinction of the blue, with catches dropping from almost 30,000 in 1960–1961 to 2,536 five years later. As was their wont, the whalers shifted to another species, the sei. In the ten whaling seasons between 1954–1955 and 1964–1965 the catch of seis went from 569 to 20,830. However, seis were much smaller than finbacks or blues and yielded considerably less oil. As a result, the number of whales being devoured by the factory ships each season held roughly steady, but production was falling off fast.

The European whalers were falling off with it. The Europeans killed whales solely for oil, rendering even the meat. Since the meat of a rorqual contains only 5 percent fat, rendering was hardly worth the trouble, but the Europeans had nothing else to do with the meat because, although it was similar in appearance and taste to beef, practically no one outside Norway would buy it. Rendering the meat, though, actually lowered the quality of the oil. Only blubber and bones went into the cookers on Japanese factories, and their product was of higher quality. The meat was frozen or salted for consumption back home in Japan. Because of the strong Japanese market for whale meat, Japanese whalers earned much more for each whale than did the Europeans. George Small calculates that the average blue whale was worth about $3,700 to the Europeans and $11,250 to the Japanese. The demise of the finback and the blue and the forced move to the sei hurt the Europeans much more than the Japanese. One sei whale produced only one-sixth as much oil as a blue whale, but it yielded almost one-third as much meat. The Japanese could float their industry on the meat of the sei, but the Europeans, who had only a smaller quantity of low-quality oil to sell, were faced with a commercial extinction as certain as the one to which their own rapacity had doomed the blue whale.

But the Japanese, too, would have finally gone under, had it not been for good economic and geographic fortune. Since the end of the Yankee era, the sperm whale had been a poorly regarded creature, hunted only here and there. In the Antarctic, sperm carcasses were lashed as fenders to the sides of factory ships to prevent collision with buoy boats and tankers and cut adrift when the reek grew powerful enough to penetrate even the factory's overpowering stench. Then in 1962, the price for sperm oil, in demand again as a fine lubricant because of its stability under high pressure and temperature, made a sharp jump. For the first time since the Yankee boom, killing sperm whales in a big way could bring big money.

This development worked largely to the benefit of Japan and the Soviet Union. A large number of sperm whales was to be found in the North Pacific, where the preexploitation population was about 300,000 whales. Also, the North Pacific contained attractive numbers of rorquals, on the order of 45,000 finbacks and 75,000 seis. Japan and the Soviet Union could keep at least a portion of their pelagic fleets busy most of the year: southern summer in the Antarctic and northern summer in the North Pacific. The Europeans could not afford to operate their fleets all year. Between several months' enforced idleness and the collapse of the Antarctic grounds, the European whalers went under. By 1967 Holland and England were out of the business and the North Pacific catch had surpassed the Antarctic. Only Norway remained a short while longer, mostly by killing minke whales and selling the meat to Japan.

Since the IWC set no quota on sperm whales or on rorquals outside the Antarctic, fleets whaling the North Pacific were bound only by size limits and by the protection of calves, females with calves, and grays and rights. As early as 1964, the Scientific Committee suggested regulation for the area, but nothing happened until 1966, when Japan, the Soviet Union, the United States, and Canada met. The humpback and the blue were protected in the North Pacific, a move which, like all protections till then, was behindhand. According to the best estimates, the blue whale population in the whole vast North Pacific did not exceed 4,900. The animal was too rare to hunt profitably. As for the humpback, it was already commercially extinct. One of the last local populations, some 900 humpbacks along the central California coast, had been reduced to but a few dozen animals by catcher boats working from two shore stations in San Francisco Bay. The four-nation meeting set a quota on sperm whales, but it was voluntary. No agreement could be reached on the rorquals in 1966 or at a similar meeting in 1967. The IWC set limits on the take of finbacks and seis in the North Pacific in 1968.

Prior to that same IWC meeting, the Scientific Committee gathered to talk about age determination in cetaceans. The discussion was not so eso-

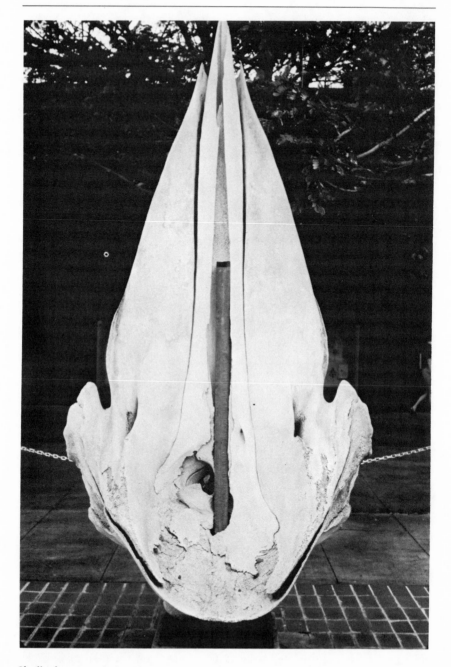

Skull of sperm whale (*Physeter macrocephalus*)

teric as the topic sounds. Age determination is central to the complex bio-statistics by which wild populations are censused and their reproductive potential is assessed. The scientists at the meeting agreed among them-selves that previous methods of determining age were inaccurate and that new research on earplug laminations indicated smaller populations of whales than had generally been assumed. As a result, the Scientific Com-mittee proposed that the Antarctic quota be cut from 3,200 BWU to 2,700. Japan and the Soviet Union refused to comply with the recommen-dation, and had not Norway come to the rescue, the IWC might once again have come apart. The Norwegians indicated that they planned to keep their pelagic fleet home that year but that they wished to retain their quota in the hope of returning to the Antarctic the following season. Thus the quota was set at 3,200 BWU but since the Norwegian portion was unused, the effective quota was 2,700. Norway never made its planned return to catching large rorquals, and in both 1969–1970 and 1970–1971 the quota was set at 2,700 BWU.

With the withdrawal of Norway from pelagic whaling the makeup of the IWC had changed. In the 1950s, with Britain, Russia, Japan, Holland, Norway, and even Panama in the Antarctic, the whalers had their way. It took a three-quarters majority to change the Schedule, and even the most determined conservationist position could make no parliamentary head-way. With Norway gone, whalers were a minority of two. The voting arithmetic of the IWC added up to a new total.

Meanwhile, opposition to whaling was mounting steadily in the United States, partly because of John Lilly's ruminations on dolphins and Roger Payne's recording of humpback songs and partly because of an ever-widening politics of the environment. In late 1970 Secretary of the Interior Walter J. Hickel added eight species of whale to the list protected by the Endangered Species Act. The listed species represented the five protected by the IWC — blue, humpback, bowhead, right, and gray — plus, surpris-ingly, three still hunted from the one remaining United States shore sta-tion — finback, sei, and sperm. Representatives of the Del Monte Fishing Company, which operated that station, maintained that the three hunted species had populations large enough to sustain continued slaughter. Hickel agreed with the company's numbers but not its optimism. "It is . . . clear," he said, "that if the present rate of commercial exploitation continues unchecked, these three species will become as rare of the other five. We are not going to wait until all these species are on the brink of extinction before we take positive action." The Del Monte Company was allowed one final season before the government canceled its license to whale.

Official attention moved to the international area. Politicking in the IWC produced the beginnings of a conservationist coalition. Mexico and Argentina stood fast with the United States. France, Canada, and Britain joined with them from time to time.

The first small victory came in 1971. Enforcement was a weak point in the IWC charter. Policed only by government inspectors, factory vessels and shore stations were only as careful in their observance of rules and regulations as their governments forced them to be. The Norwegians and the British tended to be scrupulous, more so, it seems, than the Russians or Japanese. The worst record belonged to the Panamanian factory vessel *Olympic Challenger,* which, according to affidavits from disaffected crewmen, had falsified catch reports and killed whales without regard for size, sex, species, or season, all with the connivance of the putative inspectors. One way to at least reduce the number of such violations was to have the inspectors hired and paid by a government other than the one licensing the whaling operation they were to inspect. The idea had been talked about since the 1950s, but no specific plan made its way through the whalers' opposition till 1971. That arrangement, called the International Observer Scheme, provided for reciprocal exchanges of inspectors between member nations and has been in full operation since the 1972–1973 season.

An even bigger victory came with the dismantling of the BWU system. The BWU had always made scientific nonsense, but the whalers had resisted more rational species quotas because such limits would almost certainly cut into their catches. Antarctic whalers did not always know which kind of whale they would find where. Since oceanographic conditions, like the weather, varied from year to year, the feeding grounds likewise shifted around. The BWU allowed the whalers to kill what they found regardless of species. Quotas set species by species would make life more chancy. A fleet killing finbacks would have to turn to seis once the finback quota was filled, but if luck ran bad and no seis were found, then the take would drop even if finbacks could still be had. As much as the whalers disliked species quotas, though, the destruction permitted by the BWU system had become all too apparent. At the 1971 meeting a motion to drop the BWU from the Schedule passed, and at the 1972 meeting separate quotas were set for finbacks and seis, with the latter species defined to include Bryde's whales as well. Minke whales, although not counted under the BWU system and therefore taken at will, were also granted a quota even though, in the words of the IWC's then chairman, J. L. McHugh, "this resource is under-harvested." Sperm whale quotas, already set separately for each sex in the North Pacific, were extended to the Southern Hemisphere for the first time.

The United States and its allies were prepared to go much farther, to strike at the whalers' very right to whale. In 1972 the House of Representatives and the Senate passed resolutions calling for a ten-year moratorium on commercial whaling. At the United States' prompting, a resolution recommending such a moratorium was added to the agenda of the upcoming United Nations Conference on the Human Environment, scheduled to meet in Stockholm in June only two weeks before the IWC got together in London. President Nixon appointed Russell Train, chairman of the Council on Environmental Quality, as his special representative to the IWC, announcing that this appointment demonstrated the high priority his administration placed on the conservation of the world's whales. Robert M. White, head of the National Oceanic and Atmospheric Administration (NOAA), led the American delegation to the Stockholm conference, a delegation including among its members the governor of Georgia, Jimmy Carter. White delivered a statement calling for an amendment to strengthen the resolution supporting the whaling moratorium. The amended resolution read: "It is recommended that Governments agree to strengthen the International Whaling Commission, to increase international research efforts, and as a matter of urgency, to call for an international agreement under the auspices of the IWC and involving all governments concerned, for a ten-year moratorium on commercial whaling." White's statement drew a standing ovation and strong opposition from Japan. The amendment passed 51 to 3 (Japan, Portugal, and South Africa), with 12 abstentions, and emerged from the plenary session without a single negative note. Buoyed by this impressive mandate, Train swept into the IWC meeting with the moratorium resolution. The reception was not exactly kind; someone was heard to mutter, "Stockholm was an expression of opinion; here we are in the real world." Even the Scientific Committee, supposed guardian of rationality and objectivity, opposed the resolution. The committee argued that a blanket moratorium could not be justified because "prudent management required regulation of the stocks individually." Perhaps, because it had taken so long to drop the BWU for species quotas, the scientists were wisely chary of the political aftermath of supporting the moratorium. But the scientists didn't stop there. They went on to argue that an end to whaling would also mean a lamentable end to whale research — and, presumably, to some of their jobs. The scientists were actually saying that the whales had to be slaughtered to be understood, a logic only slightly less specious than the military commander in Vietnam who destroyed the town in order to save it. The IWC commissioners in favor of whaling seized upon these self-serving arguments just as they had earlier clasped Slijper's blind stubbornness to their commercial bosoms. Votes on the moratorium ran 4 in favor (United States, Britain, Mexico, Argentina), 6 opposed (Iceland, Japan, Norway, Panama, South Africa, Soviet Union), and 4 abstaining (Canada, Australia, Denmark, France).

In 1973 Robert M. White, now the United States' IWC commissioner, again submitted a proposal for a blanket moratorium. The Scientific Committee again responded that such a moratorium was not required biologically nor could it be justified scientifically. In plenary session the Soviet Union's Commissioner Nosov and Japan's Fujita used the Scientific Committee's arguments against the proposal, which was again defeated but by a narrower margin: 8 in favor (Canada, France, Panama, and Australia had come over), 5 opposed, and 1 abstaining (Denmark). Nosov and Fujita were something other than pleased with the vote, though, because it was obvious that the unrestrained whaling their nations conducted was fast losing support. Japan and the Soviet Union proposed an Antarctic finback quota of 1,650. The United States suggested 0. The chairman of the IWC, I. Rindal, put forward a compromise figure of 1,450 and suggested that all whaling for finbacks in the Antarctic cease within three years. The compromise passed. The Soviet Union and Japan accepted the Scientific Committee's recommended quota of 4,500 Antarctic sei and Bryde's whales, but they disliked the recommendation of 5,000 for minke whales and the proposed division of the Southern Hemisphere's sperm whales into nine separate management zones, each with its own quotas. They wanted 12,000 minkes and no division. The voting forced them to accept 5,000 and three zones. Now the two pelagic whalers turned obstructionist. Japan made noises about objecting to the minke quota. And when the Finance and Administration Committee met to discuss measures to increase the IWC's funding in order to strengthen the organization as called for by the Stockholm conference, Fujita and Nosov said that with the future of whaling in doubt, they were unprepared to consider any new funding. As the meeting adjourned, they dropped loud and dark hints about canceling the International Observer Scheme in the Antarctic.

Japan and the Soviet Union continued to exchange observers on their factory ships, but they both objected to the minke whale quota and to the division of the sperm whales. They took almost 8,000 minkes, a slaughter that struck at the whales' survival and at the IWC's viability.

The United States decided that when push comes to shove, shove back. According to the 1971 Pelly Amendment to the Fisherman's Protective Act of 1967, the President can order an embargo of the fishery products of any nation whose actions undermine the effectiveness of an international conservation organization. Since tradition counts whaling among fisheries, the Pelly Amendment applies to the IWC. The secretary of commerce certified that the objections by Japan and the Soviet Union had undermined the IWC and merited embargo. Such an embargo would hurt; the two nations export about $100 million in fish and shellfish to the U.S. each year. President Ford withheld action. He let it be known that he wanted to see whether the two nations would soften their long-standing obduracy at the upcoming IWC meeting.

That meeting opened again with Commissioner White's proposal for a ten-year moratorium on commercial whaling. The Australians put forward an amendment to that proposal which has served as the rubric of the IWC since then. Japan and the Soviet Union voted against the Australian amendment, but the message about the Pelly Amendment embargo got through and they have adhered to the new order.

The Australian plan, called the New Management Procedure (NMP), mixes old and new in a compromise that falls well short of perfection but does improve on a sorry past. Basic to the NMP is a fuzzy but favorite concept borrowed wholesale from fisheries management: Maximum Sustainable Yield, or MSY. The theory of MSY is that proper management of a population will bring it to a level that provides the largest possible annual increase, which humans can then "harvest" year after year without depleting the reproductive population. With whales, the conventional scientific wisdom — based on poor, paltry data — has been that the MSY level equals 50 percent of the unexploited population. MSY sounds scientific and considered, but it is actually neither. For one thing, MSY comes from fisheries — where, by the way, it has proved less than a success — yet whaling and fishing have little in common apart from the sea. Whales are not netted in schools, but run down and shot as individuals. The population dynamics of whales are those of mammals, not fish, and the figure of 50 percent for MSY is just to the guesswork side of arbitrary. In addition, MSY is a simplistic concept. It ignores esthetics, economics, and ecological relationships.

The saving grace of the NMP is that it applies the blunt and bad tool of MSY to a more sharply defined task. Cetologists have found that all the whales of a particular species in a given ocean cannot be lumped together. Instead, the whales have local populations called stocks, each with its own range. The stocks intermingle little. If the finbacks around the British Isles were destroyed, it is unlikely that finbacks from the East Greenland–Iceland stock would enter the emptied area and colonize it. When a stock is exterminated, it is gone for good. Thus the important biological unit in whales is the stock, not the species, and if management is to take on even the appearance of rationality, it must be by stock and stock alone.

Under the New Management Procedure, the Scientific Committee divides each species in each ocean into stocks on the basis of the best data and population models. The scientists then assess the size of each stock and classify it into one of three categories: Initial Management Stock, which exceeds the MSY level; Sustained Management Stock, which is at or near MSY; and Protection Stock, which has fallen below MSY. Quotas are set for formulas for Initial Management and Sustained Management Stocks, and no whaling of a Protection Stock is allowed until it increases enough to become a Sustained Management Stock. To prevent the destruction of a small stock before anyone knows how vulnerable it is, no hunting of Initial Management Stocks is permitted until the population has been determined. To remove considerations of profit from the setting of quotas, authority for the task rests with the Scientific Committee, which is supposed to be disinterested. Bickering and infighting have continued under the NMP, but the IWC's members have adhered to the Scientific Committee's recommendations. Neither Japan nor the Soviet Union has filed an objection since 1973.

The NMP has made for definite improvement if only because it takes the continued existence of each species as its paramount value. Blue whales and humpbacks are now protected in all oceans, with the single exception of a subsistence kill of ten humpbacks allowed the Greenlanders. Finbacks are protected everywhere except the North Atlantic. Seis and Bryde's have been found to occupy separate ranges in the Antarctic. The Bryde's have been censused accurately in only one area of the Antarctic and only there is whaling for them allowed; elsewhere they are classified as an Initial Management Stock and protected. Seis are protected in the Antarctic now, along the eastern coast of Canada, and in the North Pacific. Sperm whales in the Southern Hemisphere, which have been badly overhunted in the past few years, are now protected in most of their range, with all whaling scheduled to end in 1982. Total quotas have declined steadily: from 37,300 in 1974–1975 to 14,553 for 1980–1981.

There have been other gains. Some whaling nations that had long remained outside the IWC — namely Spain, Chile, Peru, and South Korea — have joined, bringing their previously unregulated catches under the quota system. And at the 1979 meeting the IWC made its two greatest strides toward preserving the world's whales since dropping the BWU and adopting the NMP. First, the IWC made the entire Indian Ocean north of 55° south a whale sanctuary. No longer may whaling vessels or shore stations operate in these waters. Second, the IWC banned all factory ship whaling, except for taking minke whales in the Antarctic.

It is, however, easy to put too optimistic a face on these developments. Quotas have gone down because whale stocks, for all the polysyllabic scientific talk that fills IWC meetings, have continued to decline. There are fewer whales left in the seas to catch, and the lowered quotas reflect continuing destruction more than increasing restraint. As for the closing of the Indian Ocean, it was largely symbolic. The area was a minor hunting ground and the prohibition on whaling had little financial impact. The ban on factory-ship whaling simply moved up the date of the inevitable. Pelagic whaling is a high-capital, high-fuel enterprise, and between failing catches and rising oil prices, profits have grown slim. Even before the announcement of the ban on pelagic whaling, the Soviet Union was letting its Pacific whaling fleets rust, and now the factory vessels based in Vladivostok have been converted to fish processing. Only the fleets based in the Baltic and Black seas remain for Antarctic whaling. The Japanese, too, have hit upon hard times. Some years ago the three major whaling companies consolidated their operations in a joint corporation and laid off half their employees to cut costs. The big-money days of large-scale whaling are past.

On the down side, there has been a subtle yet disturbing change in the proceedings of the past few IWC meetings. The key to the NMP is adherence to the quotas recommended by the Scientific Committee. This ideal held in the NMP's first years. But of late the IWC's way of setting quotas has looked more like something reminiscent of the backroom haggling of the 1950s and 1960s. And the author of that change, sadly, is the United States.

The change came when the United States found itself on the receiving end of one of the Scientific Committee's conservation decisions. At issue was the bowhead whale. The bowhead is protected, but the Eskimos, under the aboriginal exemption of the IWC charter, could take the animals for their own use. Bowhead catches by the Eskimos were low, both because the species had never recovered from the Yankee ravages and because there were only so many whaling crews. Whaling gear cost about $9,000 and it ran a whaling captain another $2,000 to feed and shelter his crew for the spring season. Few men ever accumulated the necessary wealth, and most captains got their gear by inheritance or marriage.

Then native land claims were settled and the building of the Alaska pipeline provided jobs. Eskimos put some of their new money into whaling gear, and the number of crews almost doubled. The catch rose too, from an average of only 9 between 1900 and 1969 to 48 in 1976. Also, with inexperienced whalemen on the water, more whales were hurt but not captured. In 1977, as the worst example, 29 whales were taken, but 84 escaped after being harpooned or bomb-lanced. Some of the animals no doubt survived, but others surely died later.

These numbers worried the statistically minded types who make up the Scientific Committee. They knew that the bowhead, which they thought totaled only about 1,000 animals, could not long sustain such a kill. So, quite to the surprise of the United States delegation, the committee recommended to the 1977 IWC meeting that the Eskimos' aboriginal exemption be lifted and the bowhead hunt be banned. The conservationist nations — seeing the obvious logic behind the recommendation — agreed with the committee, and the whaling nations — eager to see the United States get the sharp end of the stick for a change — went along with a will. The resolution to ban the bowhead hunt passed unanimously, with the United States abstaining.

The Eskimos, understandably, were furious. The bowhead hunt is central to their sense of community and the meat is important in their diet. They threatened to hunt as they wished and the IWC be damned. The American government was caught between an Eskimo rock and an IWC hard place.

Seeking a way out of the dilemma, the United States found unlikely allies: Japan and the USSR. At the same meeting where the United States was hit with the bowhead ban, Japan and Russia watched the North Pacific sperm whale quota drop from 7,200 to 763, supposedly because of a new computer program. Japan and the USSR asked for a special meeting in December to work out the computer problems and arrive at "correct" quotas. The United States went to that same meeting in Tokyo to negotiate the bowhead problem. The talks were conducted behind closed doors. When the discussions ended and the doors opened, Japan and the USSR had a sperm whale quota of 6,444 and the United States had a bowhead kill of 12 captured or 18 struck, whichever came first. The principals all deny a trade-off, but the quid pro quo looks too pat to be coincidental.

The United States went into the June 1978 IWC meeting ready to argue for an even bigger bowhead quota. During the spring the government had spent over $700,000 to send a group of scientists to northern Alaska to make a census of the bowheads. They found more than had been thought to exist, about 2,300, according to the best guess. The Scientific Committee, however, again recommended that no bowheads be taken, but the United States lobbied for and won a quota of 18 killed or 27 struck. Much the same scene was repeated in 1979 and again in 1980.

The United States' position in the IWC has changed. Before, the American delegation led the conservationist wing and held to uncompromising implementation of the NMP and severe cuts in the quotas. Now the United States goes into the IWC meeting with the bowhead quota as its first priority, pitting itself against the Scientific Committee and bargaining with other whalers to gain their votes on the bowhead kill. Of course, the United States has to return the favor. So, in 1979, when the Scientific Committee recommended a catch of 153 Bryde's whales off western South America, the United States joined with Peru, Chile, and Japan to vote in a quota of 264. Once more the NMP bit the dust.

The IWC is a weak organization, hamstrung by its own internal politics and lacking the punch and power to make strong rules and make them stick. Because it lacks real authority, every step the IWC takes to save the whales paradoxically endangers them. With quotas dropping and whales becoming scarcer, it grows increasingly attractive for the whalers to work around the IWC and take all the whales they can while the animals last. They can politick, following the United States' example. They can hunt species not covered by IWC regulations; thus the Soviet Union has been taking big catches of killer whales to make up for the slashed quotas of sperms. They can import whale products from nonmember nations. Or they can set up whaling operations completely outside the IWC. These are the pirate whalers, vessels of uncertain ownership and clouded registry that whale as they wish and sell wherever they can — usually to Japan. The most notorious pirate whaler was the *Sierra*, which worked under the cloaking mixed nationality of a ring of gun-runners, freebooters, or mercenaries. Ownership of the *Sierra* was held by a South African through a corporation chartered in Liechtenstein, the flag was Cypriot, the master was Norwegian, and the label on the ship's frozen whale meat read "Produce of Spain."

The *Sierra* killed with a cruelty appalling even in the brutal whaling business. To save as much meat as possible, the *Sierra* used a barbed metal harpoon without an explosive grenade. Struck whales commonly took hours to die, bleeding slowly to death and disemboweling themselves in their struggles against the harpoon. When the whale died, often only the prime cuts were taken, meaning that a 40- or 50-ton animal died for the sake of 2 or 3 tons of meat. The *Sierra*, fortunately, no longer sails. An explosion sent the ship to the bottom of Lisbon harbor in July 1979. But there are even now more pirate whalers at sea following in the *Sierra*'s bloody and destructive wake.

Talk of politics, parliamentary rubrics, loopholes, and pirates begs the real issue: should commercial whaling continue?

The IWC, for all its talk of scientific management, has yet to show that it can manage even one stock scientifically. Whaling has meant destruction; scientific whaling has meant slow destruction. The difference is pace, not result. Management may in fact be impossible, for our ignorance of cetacean biology is great. Yet cetologists claim still that they can control populations of creatures whose ways they barely fathom. Few conceits are bolder.

But even if the whales could afford biologically to give up the number of carcasses the scientists claim they can, should they?

Whaling is a small industry. It employs only a few thousand people worldwide and brings in annual revenues of perhaps $100 to $150 million. If the enterprise disappeared from the face of the earth this very moment, the world economy would not even notice and no national economy would suffer anything but tiny, temporary inconvenience.

Earnings, of course, tell us only market values; they say nothing about intrinsic worth. If whales gave us unique and irreplaceable commodities, whaling could perhaps be justified.

Until recently, the one unique cetacean product was sperm oil, which because of its stability is used as a high-grade lubricant, particularly for fine machinery under conditions of extreme temperature and pressure. The Soviet Union's interest in sperm whales springs at least somewhat from its use of sperm oil in strategic nuclear missiles. No longer, though, is the sperm whale the sole supplier of such oil. The nuts of a shrub of the American and Mexican deserts called jojoba or goose nut (*Simmondsia chinensis*) produce an oil so similar to sperm that only a chemical specialist can tell them apart. In fact, jojoba is somewhat superior to sperm. It has none of the gamy odor of the sperm oil, and it comes from the plant so pure that it need not be refined as sperm oil must be. The present supply of jojoba is limited, but commercial development has begun in both the United States and Mexico. In a matter of years jojoba can fill the demand for sperm oil and eliminate that reason for killing whales.

But, Japanese whalers argue, there is more to whaling than lubricants for industry. Spokesmen and apologists for the industry — from former Prime Minister Takeo Fukuda to the Japan Whaling Association's publicity puffs — maintain that Japan whales because its people need whale meat to enrich their protein-poor diet.

In fact, whale meat has become less and less important in Japan. Back in the early 1960s whale accounted for almost one-third of Japan's meat supply. The low price of whale made it an attractive substitute for pork and poultry. But since then, the price of whale has risen while that for pork and poultry has dropped. Now whale makes up only 6 percent of

Japan's meat, and the typical family eats but 3.4 pounds of it a year. Whale is a minor entrée on Japan's tables. Nor is the protein it supplies crucial. If it were, how does Japan manage to export 250,000 tons of seafood a year?

Since the beginning of World War II, 1,540,000 whales have been reported killed, bringing the total since 1900 to 2,234,000. More whales have been destroyed in our time than in any other, and they have all died stupidly, to serve purposes better met by vegetable oils, jojoba, and beef cattle. "In the world of mammals," writes Teizo Ozawa, "there are two mountain peaks. One is Mount Homo Sapiens, and the other Mount Cetacea." How odd that the one pinnacle has chosen to lay the other waste so senselessly.

---

and Venus among the fishes skips and is a she-dolphin
she is the gay, delighted porpoise sporting with love and the
* sea*
she is the female tunny-fish, round and happy among the
* males*
and dense with happy blood, dark rainbow bliss in the sea.
— D .   H .   L A W R E N C E

The Tuna-Dolphin Tangle

T he tunamen never had anything against the dolphins. They were tunamen, and taking tuna, not killing dolphins, was their trade. But the dolphins ran afoul of the gear and died in the net and that was just how it was. The tunamen threw the dead dolphins overboard with speed and without sentiment. They had to get the fish into the holds and ready the net for the next set and push on for more tuna. The season was too short and the note on the boat too big to diddle around. They battened down and were off, and in their wake floated the dolphins, snouts upright in death.

It has not always been this way. In fact, tuna fishing itself is something of a novelty. The giant bluefin tuna, which reaches an adult weight of about half a ton, has been hunted by harpoon along the Atlantic coast for more than a century, but this was a small industry serving local markets. The bluefin gained status when anglers discovered that it would take a hook. The fish's enormous size and its incredible running speed of forty knots made it an impressive game fish and an obsession for some sports-men, among them Zane Grey and Ernest Hemingway. Yet the bluefin was not all that commercially desirable in this country. The Japanese relished the fish because its rich red flesh sliced thin and served with a piquant sauce makes the very finest *sashimi*. Americans, of course, are squeamish about raw fish, but the discovery that tuna could be canned and mixed with mayonnaise for sandwiches or salads or cooked into cheap casseroles prompted the growth of tuna fishing in this country. The market for canned tuna grew markedly after the discovery that smaller tunas, particularly the yellowfin, skipjack, and albacore, all of them lighter-fleshed than the big bluefin, canned even better and pleased the American palate more. Tunafish fast became a cheap protein staple of the American diet, and the tuna fleet grew and prospered.

The tuna boats developed several methods of taking tuna, depending on the species and the area and conditions of the fishing. Pole and line became the principal way of catching yellowfin, particularly among the boats that ventured farther and farther from San Diego and San Pedro down the coast of Mexico and into Central and South America. Chumming the water with baitfish brought the yellowfin in close to the vessel and put them into a feeding frenzy. The fishermen then fished with strong bamboo poles and bare barbless hooks, which the tuna in their fervid feeding struck as if they were baited. The fishermen muscled the impaled fish aboard, flipped them into a waiting well, and dropped the flashing hooks into the water again.

As early as 1914 the tunamen experimented with nets, namely purse seines of a design adapted from the salmon purse seines of the Pacific Northwest. The typical early purse seine ran 1,200 feet long and about 100 feet deep. The net was set around a school of fish, suspended from its floating corkline something like a hanging curtain, then its bottom was cinched up, much like the top of a woman's drawstring purse. Netting, though, had only limited success, largely because cotton twine decayed fast in warm water. By the 1930s most of the purse-seiners in Southern California shifted to mackerel and sardines, pursuing tuna only in the off-season. They stayed close to their home ports. The pole-and-line baitboats pushed farther and farther south into the eastern Pacific. Essentially there were two tuna fleets: a local purse-seine fleet and a long-distance baitboat fleet. In 1950 the baitboats outnumbered the purse-seiners 204 to 67.

Then in the 1950s an economic squeeze caught the tuna industry hard. As part of its economic rebuilding program after the war, Japan was developing its fishing industry into a global power. The Japanese fished differently from the Americans, with a method called long-lining they still use widely. Lengths of quarter-inch rope suspended from floats are laid out over as much as fifty miles of ocean. From the ropes hang thousands of lines, each with a baited hook. The captured tuna are then processed aboard a mother ship that accompanies the fleet of long-liners. This system worked with such efficiency that Japanese tuna could undersell American tuna, and imported fish was fast pushing the domestic product aside. The United States tuna industry had to improve its operation quickly, or it was doomed.

The Japanese tuna industry was labor-intensive. The Americans, aided by two key technological innovations, went the capital-intensive route. The first change was an all-nylon net, first used by a Peruvian boat in 1954 and soon adopted by American vessels. Nylon did not decay in warm tropical water like cotton, thus requiring only a fraction of the maintenance of the old gear. But even more important was the powerblock, which brought the net aboard much faster after each set and thus allowed that many more sets and that many more tons of fish each day. And now the two American tuna fleets got together. The purse-seiners wedded their improved technology to the baitboats' knowledge of the fishing grounds of the eastern tropical Pacific.

This change to capital intensity paralleled the huge postwar investment of capital in Antarctic whaling — producing the same financial exigencies and the same instrumental illogic. By 1961 the number of purse-seiners doubled and the baitboats declined. At first, since the hulls did not have to be altered, baitboats were converted by clearing away all the bait gear and replacing it with the masts, booms, winches, davits, turntables, powerblocks, and fish hoppers needed in purse-seining. Conversion was hardly cheap. It cost from $50,000 to $100,000 plus $50,000 for the net. And the investment continued. Some of the converted baitboats were replaced by newly built custom seiners of markedly larger size. Before 1961 the largest seiners could carry just under 600 tons, and the average seiner's cargo capacity lay somewhere between 180 and 275 tons. The 38 seiners added during the 1960s averaged 655 tons. Since then a few superseiners of 1,500 to 2,000 tons have been put in the water. In the 1960s alone, the United States gained some 25,000 tons of cargo capacity in its tuna fleet.

The effect of all this was to create an anomaly in American fishing. Back about the turn of the century, the United States, which had been a maritime power, abandoned the sea, except as a theater of war, for its own west. Shipping and fishing steadily declined. Today, as sophisticated and technically advanced fishing fleets from the Soviet Union, Japan, Poland, and East Germany strip the seas with the thoroughness our petroleum companies devoted to Texas, American fishermen ply their trade in almost traditional fashion from small coasting boats. The exception is the tuna fleet. The approximately 140 tuna purse-seiners make up 70 percent of the total tuna-fishing capacity in the eastern Pacific, which is the principal ground for yellowfin. These tuna clippers have the best of electronic gear, the finest crew quarters — including carpeting and paneling and even piped-in stereo sound — and a few carry a helicopter or a seaplane for fish spotting. And the tuna boats are not simply functional. They are lovely vessels, sleek and fast, all rake and good lines. They look more like scaled-up yachts than working boats, vessels of beauty rare in an age whose maritime tastes have been vulgarized by squat container ships and the bulbous monsters called supertankers.

Of course, such luxury and design cost plenty. The *Elizabeth C.J.*, for example, a 1,700-ton superseiner launched in the spring of 1976, ran her investors $5 million. The only way the *Elizabeth C.J.* and all the other seiners and superseiners can make back their staggering investments is by catching tuna, lots and lots of tuna. And that is where the dolphins figure in.

To catch tuna you first have to find them. Along the Central and South American shores the tuna school and feed close to the surface, and a good man with binoculars in the crow's nest can spot the fish themselves. Both inshore and out, the purse-seiners also look for floating debris like timber or hatch covers and set their nets around it. They call this log-fishing, and it works because the tuna often gather under such objects. During the years when the baitboats were being converted to seiners, the tunamen found that dolphins — which they call porpoises — were even better to look for than logs. Pole-and-line fishermen often saw dolphins mingling with the yellowfin they chummed, but they just assumed that the dolphins were coming in for a free meal. But as seining became more widespread, the fishermen found that the dolphins weren't simply late-arriving opportunists. Instead, the schools of tuna in the area of the Pacific from Cape San Lucas in Baja California to Chile and offshore as far as Clipperton Island and beyond associated with schools of dolphins. Once that association was formed, the fish would not forsake the mammals but would follow them wherever they went. No one knows why the tuna establish this curious link. Some scientists think the tuna may feed with the dolphins, and stomach analyses do show some of the same foods, but other researchers believe that the tuna use the dolphins for orientation in the unmarked open sea. The fishermen don't care why. They care simply that it is and that it helps them catch tuna. That is, after all, what they are about. They call this technique "fishing on porpoise." It accounts for about half the yellowfin taken every year, which in turn accounts for about half the annual U.S. catch of tuna.

The animals the fishermen seek are principally two species of the dolphin genus *Stenella*, which the fishermen dub the spinner and the spotter. While on the tuna grounds, the purse-seiner crew mounts a dolphin watch through the daylight hours, with two men on the bridge using high-power rack-mounted binoculars and another on the crow's nest using handheld glasses. They scan the horizon ceaselessly, seeking sign of dolphins or of birds like terns and boobies feeding over them. At the sighting of a dolphin school, the purse-seiner runs over for a close look. As the boat approaches, the crow's nest spotter looks for tuna signs: feeding on the surface, flashes of light reflecting off fish below, or the black spot of a dense school beneath the surface. Even if there are no signs of fish, the captain sometimes decides to gamble and set anyway.

Now the action begins. The fishing captain, who may or may not be the same as the boat captain, goes up the mast with his radio as crewmen lower two or more speedboats into the water. These boats are about 16 feet long and carry big 100-horsepower outboards that push them over the swells at such speeds that the drivers must wear safety belts and crash helmets. The skiffs take off after the dolphins. Acting on orders radioed from the fishing captain in his vantage point on the mast, the boats herd the dolphins together and chase them into position for putting out the net. When boat and dolphins are in the right place, a crewman on board knocks out a pelican hook with his sledgehammer and the net skiff, attached to the forward end of the purse seine, goes overboard. The skiff acts as a sea anchor, holding the net as the purse-seiner pulls away and the mile-long net pays out over the stern. The seiner lays the net in a circle surrounding the subdued and exhausted dolphins and the tuna, which have followed the dolphins and now school beneath them. With the net circle closed, the net is tied off on the vessel's port side, and the cable that purses the bottom, now hanging 250 feet into the water, is attached to a winch on the seiner and drawn up. This purses the net, enclosing the fish and the dolphins both. Once the cable is drawn tight, the end of the net is run into a powerblock overhead, and retrieval of the net and the catch begins.

In the early days, what followed was for the dolphins an inexorable disaster. With their sensitive sonar the dolphins knew the net was coming in upon them. A few dolphins, males usually, might jump over the corkline and escape, but the pregnant females could not and the females with calves would not. Pushed into a tiny area by the closing net, the dolphins finally panicked. The highly strung spinners commonly went into catatonia and sank to the bottom of the net. Others rushed at the barrier to break through and only caught themselves by beak or flippers or flukes. Unable to rise to the surface and breathe, the snared dolphins smothered. The final tightening of the net into a bag against the side of the ship, called sacking up, crushed many of the remaining animals, killing them outright

Spinner dolphin and calf (*Stenella longirostris*)

or leaving them grotesquely and mortally injured. Then the contents of the net were scooped up onto the deck with a brail, and crewmen sorted the haul. The tuna went into refrigerated holds, the dolphins, living and dead, over the side. Sometimes sharks stood by to grab a quick snack.

Sorting and dumping all those dead dolphins slowed fishing down, and the name of the game was getting the most fish the fastest. In 1958 Captain Anton Misetich of San Pedro worked out a way of separating dolphins from tuna before the net was sacked up, and by 1961 practically all the purse-seiners were using Misetich's technique. When about three-quarters of the net had been retrieved, the vessel maneuvered so that the portion of the net farthest from it shaped itself into a long thumb. The dolphins congregated in the thumb while the tuna swam back and forth from the dolphins to the seiner. As the tuna approached the vessel, the captain gunned the engines in reverse, pulling the net's corkline six or eight feet below the surface and out from under the dolphins. This procedure was called backing down, and the net could be backed down again and again until all or most of the dolphins had been spilled out into the open sea. If everything went right, a skillful captain could rid the net of up to 95 percent of the trapped dolphins. But some of the captains were ham-handed and things often went wrong even for the skilled masters, with the wind or the sea doing something strange and collapsing the net in huge suffocating billows about the dolphins.

Backing down was an improvement, but it simply made a catastrophe into an ordinary disaster. According to the latest and best estimate, tuna fishing in the years 1959 through 1965 killed 2,600,000 dolphins.

Not that the fishermen were saying anything to anybody about it. Many of the tuna fishermen in San Diego and San Pedro are Portuguese. They came here poor and they have worked their way out of poverty with sweat and guts and grit. They give no charity and they ask none. They are hard, proud, narrow men, and they keep their business to themselves.

But Kenneth Norris, southern California's Renaissance man of biology, was studying dolphins at the time and he found that tuna boats were a good source of fresh carcasses. Norris wondered why. He interested one of his students, William Perrin, in the problem. In 1966 Perrin got himself onto a tunaboat that took about 300 tons of yellowfin and killed about 2,000 dolphins in the process. In 1968 Perrin went out again on a different purse-seiner, kept careful records, and counted 1,359 spotter and 338 spinner dolphins killed in the course of netting 312 tons of yellowfin. Every set of the seine on dolphins killed about 20 percent of the school.

Perrin presented his findings the following year at the Sixth Annual Conference on Biological Sonar and Diving Mammals sponsored by the Stanford Research Institute. Perrin's report was most restrained. He pointed out that, estimated from his 1968 voyage, tuna fishing was killing about 250,000 dolphins a year. (Actually, Perrin had seen a good cruise; totals in the mid and late 1960s usually ran closer to 390,000.) Perrin considered this slaughter economically shortsighted. If the dolphins were exterminated, the tunamen, who were taking some $12 or $13 million of yellowfin by porpoise fishing each year, would lose that portion of their total catch, and the world would lose a resource of great intrinsic value as well. And, Perrin argued, there was every reason to believe that the current rate of predation by fishermen would indeed kill off all the spinners and spotters in time. Destroying 20 percent of each school at each set was, given the low level of reproduction among mammals and the repeated sets made on the same schools in any one season, a prescription for extermination. And far more females and young died in the nets than males, a selection that should eventually make the dolphin populations demographically distorted. In the Black Sea, where the Soviet Union had used nets to capture common dolphins for human consumption, females and young also died out of proportion to their numbers. The fishery utterly collapsed inside two decades. Perrin feared the same occurrence in the eastern tropical Pacific.

Now, looking back, it seems amazing that Perrin's report caused no real reaction, particularly among the general public. But the times were wrong. The United States was still nominally a whaling nation, and only gradually was the country taking a stand against that stupid and outmoded industry. Few people really understood the desperate plight of the great whales, much less the threat to the much more numerous dolphins. Vietnam obsessed the collective mind. Amid all the maiming, napalming, and shooting, dolphins dying in tuna nets seemed so tame, so far away.

But the war cooled down, the United States campaign in the International Whaling Commission gathered force and form, and the time arrived to deal with the tuna-dolphin problem. The occasion was the drafting and passing of the 1972 Marine Mammal Protection Act (MMPA), a piece of landmark environmental legislation. The law set as its purpose the preservation of the optimum population of all marine mammals, paying particular attention to those species and stocks that were in danger of extinction. "There shall be a moratorium," the law stated, "on the taking and importation of marine mammals . . . , during which time no permit may be issued for the taking of any marine mammal. . . ." Given that *take* meant any attempt to capture, kill, or harass, MMPA apparently forbade fishing on porpoise. But the law did allow of exceptions.

Spokesmen from the American Tunaboat Association talked to key legislators during hearings on the law. Clearly, the bill as first written would cripple the tuna industry. But, the industry argued, it already had the problem in hand. All it needed was time to improve technique and increase the rescue of dolphins from the nets. The industry provided evidence of its good intentions. In 1971 Captain Harold Medina replaced the 4¼-inch mesh in the back-down area of his purse seine with a strip of 2-inch mesh. Medina's idea was to make it harder for the dolphins to plunge their beaks into the mesh and get stuck. Medina fished successfully with the Medina panel, as it came to be called, cutting the kill of dolphins by at least one-third. Seeing Medina's success, other captains adopted the panel voluntarily. Because of the tunamen's own efforts to improve their gear, Congress gave the industry a two-year grace period from the general moratorium. But Congress also made it clear that the extension was merely temporary and that the tuna industry would finally have to toe the same hard protectionist line as everyone else. "In any event," says the section of the law granting the grace period, "it shall be the immediate goal that the incidental kill or incidental serious injury of marine mammals permitted in the course of commercial fishing operation be reduced to insignificant levels approaching a zero mortality and serious injury rate."

About two-thirds of all the purse-seiners installed Medina panels by the end of 1973, and the kill dropped from 410,992 in 1972 to 173,704 in 1974. Then, even though the catch of yellowfin was virtually the same in 1975, the toll rose to 194,000. The tunamen were losing ground and falling well shy of the "zero mortality" the MMPA called for. The National Marine Fisheries Service, charged with enforcing that law, seemed little concerned, however. Perrin and other scientists were working to improve gear, but the agency had yet to make a thorough effort to census the affected dolphin populations or to determine the harm done by the killing of at least 5 million dolphins in one ocean region in little more than a decade. Nor did the fast approaching end of the tunamen's grace period impress a sense of urgency or importance on NMFS. In September 1974, the month before the end of the grace period, NMFS published the regulations tunamen would have to follow to get a permit to fish. Despite the industry's mixed performance, the rules required only the installation of a Medina panel, a few other gear changes, and a promise to back the seine down before sacking up — basically, the status quo. The American Tunaboat Association applied for a general permit to cover all its members.

The Environmental Defense Fund filed suit in Washington, D.C., to stop NMFS from granting that permit, arguing that the lax rules violated the MMPA. The suit went the way of all legal paper into the maw of the federal court system, and the tunamen got their general permit.

Meanwhile, another drama opened elsewhere. At the time, Stan Minasian, a native San Franciscan, was a student at the University of California, across the bay in Berkeley. Minasian got interested in the tuna-dolphin issue after reading a few short magazine articles about it. He wrote various people for more information and discovered that although the facts and figures were to be had, no one person had put them all together. Minasian became that person. Soon he was supplying information to others; television stations and newsmagazines called him to find out the latest on the issue. The enterprise became so engrossing that Minasian left school and his part-time job to work solely on the dolphin issue. Along the way he set up a nonprofit corporation called Save the Dolphins.

The government had in its archives footage of dolphins being trapped, smothered, and thrown overboard from the purse-seiner *Queen Mary*. The film was political dynamite, and its existence remained studiously unannounced until the government tipped its own hand. In reply to a letter from Minasian asking for film shot by William Perrin, the NMFS responded that he must be after the *Queen Mary* footage.

Minasian knew nothing of this film, but he figured correctly that he might have happened across something big. Minasian said he wanted to see the footage. NMFS said no. Joseph Medina, skipper of the *Queen Mary*, said that if NMFS released the film, he would sue. Minasian approached Ralph Nader's organization, and Victor Kramer, a Nader attorney with a reputation as a legal eagle on freedom of information cases, saw a strong argument in favor of releasing the film. Kramer contacted the prestigious San Francisco law firm of McKenna, Fitting, and Finch, which did not act as Minasian's official legal representative but did provide support services and a junior partner named Charles Miller. Minasian sued the government for release of the *Queen Mary* footage. Joseph Medina, true to his word, sued. And the American Tunaboat Association added a third suit to the pot.

After the usual long wait, the three suits were heard in federal district court before Judge Charles Renfrew. At first, Minasian says, Renfrew sided with the industry, but the continuous stonewalling by the tunamen got to him. It seemed that whenever the industry needed information from Medina, they got it, but if Minasian or the government desired the captain's services, he was out to sea and could not be reached. Minasian thinks that the tunaboat owners, ignorant that Miller was working gratis, were trying to drag the case out and bleed him dry. Once they found out about Miller, Minasian says, they came quickly to agreement. Minasian got the *Queen Mary* footage, but he had to caption the film as experimental and not necessarily representative of usual fishing conditions.

By now expert at soliciting help, Minasian got more from Westinghouse Broadcasting and San Francisco television station KPIX, Ralph Nader, the Environmental Defense Fund, and Adolph Gasser, Inc., a photographic supply company. Minasian put together a short film about the tuna-dolphin problem, incorporating the *Queen Mary* footage, had it narrated by Dick Cavett, and released it in 1976 as *Last Days of the Dolphins?*

When you see the film, you know why the tunaboat owners wanted the *Queen Mary* footage suppressed. It clearly shows the dolphins snaring themselves in the net and fighting to get free. Dolphins are about our size, and you can easily feel yourself in the dolphins' place, caught in the net, struggling to surface and breathe, and knowing with each passing second and the heating fire in your lungs that death has you harder and harder in his cold hands. It is one thing to talk roundly and abstractly about 100,000 dolphins dying in tuna nets in a year. It is an altogether different and much more horrifying thing to watch a single dolphin endure its final agony.

Last Days of the Dolphins? sounded the alarm and spread the word. It stimulated increasing support for a now two-year-old boycott of lightmeat tuna organized by the Sierra Club and Project Jonah. Particularly in California the boycott took hold. The Consumer's Co-Operative stores in the San Francisco Bay Area put up signs on the shelves of canned fish advising shoppers of the killing of dolphins in taking yellowfin for lightmeat tuna and suggesting that they turn instead to canned albacore and bonito, which are not fished on dolphins.

The government was doing little about this fast-thickening morass. The staff of the Porpoise/Tuna Interaction Program released a report in August 1975 showing that they had made precious little headway on the admittedly difficult but absolutely central task of estimating the populations of the affected species. The report admitted little confidence in the present population data. It concluded: "There is no striking evidence that the stock [of offshore spotted and eastern spinner dolphins] is either increasing or decreasing. At present the stock is probably stable or increasing or decreasing slightly." This marvelous sentence, which delicately wedded double talk and bureaucratese, won the Washington *Star's* coveted Gobbledygook Award.

In a way, NMFS was ill prepared to handle the tuna-dolphin issue. The agency is in the business of promoting and regulating fisheries, not assessing populations of pelagic marine mammals. The NMFS's stock-in-trade is boats and nets and dollars and cents. Unsurprisingly, an economic report on the financial condition of the tuna fleet written by Phyllis D. Altroggle, an NMFS industry economist, and released in January 1976 was considerably less equivocal than the dolphin population estimates. Altroggle concluded that the fleet was in a bad way and facing a probable loss in 1976. The implication was that any cut in the allowed kill of dolphins — set at 78,000 by NMFS regulations a month earlier — would kill the tunaboats instead. The dismal science was underwriting a dismal status quo.

In the spring of 1976 the suit against the NMFS permit rules came before Judge Charles Richey of the United States District Court in Washington, D.C. On May 11 Judge Richey handed down his decision, telling the tuna industry to stop fishing on dolphins as of May 31. Richey agreed with the environmentalist arguments that NMFS had failed to live up to the requirements of NMPA, thereby voiding the general permit it had granted. The tunamen blanched, the NMFS blushed, and the environmentalists blessed their attorneys.

The tunaboat owners, who had truly made little effort to comply with the very law that granted them the exemption to kill dolphins, sought to change the law instead of their ways. Robert Leggett, chairman of the House Subcommittee on Fisheries and Wildlife Conservation and the Environment, authored an amendment to MMPA. The change would have dropped the requirement to cut the dolphin kill to zero and substituted a provision that the kill be as low as current technology allowed. Leggett took this legal monstrosity as far as the full Committee on Merchant Marine and Fisheries. Then the tunaboat industry relaxed. The appeals court to which Judge Richey's decision had been referred allowed purse-seining on dolphins to continue while it considered the case.

At the end of July, out in La Jolla, staff scientists of NMFS met at the Southwest Fisheries Center to comply with the portion of Judge Richey's order affecting them. They had to estimate the current and optimum sustainable populations of each species and determine how supposed regulations would affect the numbers. "Look at the size of the ballpark," one government scientist complained to a reporter from *Sports Illustrated*. The eastern tropical Pacific is twice the size of the continental United States. "There's millions of square miles of ocean out there. Nobody's ever found an effective way to census dolphins." But the scientists at the La Jolla meeting tried, filling page after report page with statistical arcana, methodological disquisitions, and tracts on the varying degrees of mathematical certainty. The final picture, for all its bureaucratic and scientific frosting, was neither sweet nor pretty. Tropical tuna fishing affected eleven dolphin and whale species, but the kill fell disproportionately — to the tune of 99 percent — on only three: the spinner, the spotted, and the common dolphins. The common dolphin was still lightly exploited, but of the seven identified stocks of spinner and spotted, at least one, the eastern spinner, was depleted badly and two others were headed the same vanishing way fast.

By the end of October, reports from the tuna fleet showed that the fishermen had already killed the 78,000 dolphins permitted. The fleet was ordered to stop fishing on dolphins. The tunamen challenged the order in court in San Diego. The order was upheld. They took it to court again in San Francisco. It was upheld again, and the tunamen were told to cease fishing on dolphins in three days.

The tunamen had cooked their own goose. They had shown that they could not fish for a whole season and stay within the NMFS quota. In fact, they had done much worse than first appeared the case. When all the figures from NMFS observers aboard the tunaboats were fed into the computers, tabulated, and extrapolated, the actual 1976 kill amounted to 128,350 dolphins, 50,350 animals, or two-thirds, over the quota. All that fast talk in the MMPA hearings about the miracle Medina panel and the quick solution to the dolphin problem sounded in retrospect like so much empty puffery.

The months following the order to cease fishing on dolphins provided a short course in hard heads and stiff necks. When the new year came and the NMFS had not yet released its final regulations and quota for 1977, the tuna fishermen won a temporary enjoinder against MMPA from Charles Enright, a federal judge in San Diego, to permit a temporary quota of 10,000 dolphins. The victory was short-lived. On February 4 a federal appellate court in Washington, D.C., ordered that no permits for fishing on dolphins be issued until Commerce Secretary Juanita Kreps complied fully with MMPA.

The NMFS, which had been so markedly lax for so long, became newly stringent in proposing a total quota of 29,920 dolphins allocated among species and stocks. Importantly, NMFS put a zero quota on eastern spinners, a mainstay of porpoise fishing. This proposal fell well below the quotas suggested by the Marine Mammal Commission and the Environmental Defense Fund, which were in the neighborhood of 52,000. The tunamen, for their part, wanted a 96,100 total with no species quotas. All these alternatives went before administrative law judge Frank Vanderheyden, who ended up accepting the tunamen's figure of 96,100 but allocated it by species, including an allowed kill of 6,587 eastern spinners. Dr. Robert Schoning, NMFS director, took Vanderheyden's opinion under advisement as he prepared the final regulations.

During that time, Schoning put in an appearance before a House Merchant Marine Committee meeting. Gerry Studds, member of Congress from Massachusetts, asked Schoning what happened to the killed dolphins.

"They are returned to the ecosystem," said Schoning.

The audience was heard to guffaw.

Studds continued, "Is that the government's way of saying that they are thrown overboard?"

Schoning paused. "Yes," he said.

On March 1, 1977, Schoning announced in the *Federal Register* just how many dolphins could be returned to the ecosystem legally: 59,050. The total was divided among twelve species and stocks, with no kill of eastern spinner dolphins allowed. Schoning also ruled that tuna sets on pure schools of dolphins could be made only on offshore spotted and common dolphins and that no sets could be made on any mixed school containing eastern spinner, Costa Rican spinner, or coastal spotted dolphins. He imposed various gear restrictions such as floodlights for sets made at dusk and the use of speedboats during backdown.

In response, the tunamen tied up in San Diego and refused to go out. What had them stymied was the zero quota on eastern spinners. These animals are so often mixed in with schools of other species and so hard to identify in a crowd that the captains did not see how they could fish under this prohibition without bringing one fine after another down on their already troubled heads. The American Tunaboat Association played its hole card. Foreign purse-seiners have the reputation of being much sloppier about dolphins than American vessels, with NMFS estimating that they kill two and a half times more dolphins per ton of yellowfin. The tunaboat owners let it be known that they might go someplace where the restrictions were less restrictive. The *New York Times* reported that the owners accepted purchase options for more than 20 purse-seiners from Saudi Arabian oil magnates represented by Ray McVeigh, a San Diego agent. The Saudis, McVeigh said, wanted to put the purse-seiners under the flag of convenience of "an undeveloped Indian Ocean country friendly to the United States." Then August Felando, general manager of the American Tunaboat Association, spent three days in Mexico City talking with government officials, among them President José López Portillo. The Mexicans, who have made a concerted effort to develop their marine resources, supposedly wanted to buy several seiners and work out joint ownership arrangements under Mexican colors for more. The implied threat was clear enough: give us the quota we want or we'll go elsewhere and do in the dolphins anyway.

Secretary of Commerce Kreps sought a way out of the tangle of regulations, court orders, divided authority, and mistrust by announcing that tuna captains would be prosecuted only for the intentional killing of eastern spinners, not "for the accidental taking of small numbers" unidentified in mixed schools. For a bit it looked as if the tunamen would take the Kreps statement and set sail, but at the last moment they balked.

Senator Alan Cranston of California had been putting together what was charitably called a "compromise" bill. Cranston wanted the quotas put on individual vessels, which was a very good idea, and he wanted an allowable kill of 157,000 over the next twenty months, which was very bad. The Environmental Defense Fund told Cranston that they would not accept the bill, a refusal that killed it. The tunamen, despite Cranston's urging, refused to go out. But they stayed in port only a few more days. During that time, the tunaboat owners met with a New York member of Congress, John M. Murphy, who had introduced legislation to raise the total quota to 78,000 per year and allow the killing of eastern spinners. The owners felt that Murphy could win and the fleet should set sail. The Captains and Mates Association and then the Fishermen's Union agreed. The purse-seiners left San Diego in mid-May, five and a half months after the traditional sailing date of January 1. The delay, according to the American Tunaboat Association, cost $100 million.

Murphy's bill failed, beaten largely because of a strong case against it made by William Butler of the Environmental Defense Fund. A whole round of bills was submitted and debated in committee and finally left to languish as all parties waited to see how the truncated season's fishing turned out. In late December 1977 the NMFS published new rules. The quotas for 1978, 1979, and 1980 were 51,945, 41,610, and 31,150, requiring the tunamen to make steady progress toward the "zero mortality" mandated by the Marine Mammal Protection Act. But the rules also admitted the imperfection of the real world by allowing, in addition to the quotas, an "accidental" take of protected species and stocks. The purse-seiner must also meet certain gear requirements and follow procedures designed to ensure that all the dolphins that can get out of the net do in fact get out. The boats must have a so-called superapron in their nets. This long panel of mesh webbing makes for a gradually shallowing channel in the backdown area of the seine and helps separate dolphins from fish and ease them out of the net and into the open sea. The apron also helps overcome the problem of catatonic spinners. When alarmed, these animals sometimes simply drop to the bottom of the net belly-up, seemingly dead but actually quite alive, in a kind of shallow coma brought on by some combination of fright and fatigue. The fishermen used to think these animals were dead and brail them aboard; now they know better. The apron slides out from under the catatonic dolphins during backdown, spilling them out of the seine. Should any remain behind, they can be removed from the net by two crewmen required by law to stand by, one of whom must have a snorkel and face mask for underwater rescue. Also, during backdown, two speedboats must remain close at hand and be ready to tow the net to prevent any bunching or roll-up that could trap dolphins.

The tunamen, of course, grumbled about the new rules — fishermen grumble about everything anyway — but they seemed by and large to go along with them. And good luck came their way. Tuna catches in the eastern tropical Pacific in 1978 were huge, more than making up for the 1977 losses. What's more, the kill of dolphins dropped off steeply. In 1977, despite the short season, the dolphin kill totaled 51,348, and the new regulations allotted the tunamen a kill of only 51,945 for the full-length 1978 season. The tunamen did even better than they had to; they killed but 30,593. The American Tunaboat Association was quick to call public attention to its success. And the attitude of the fishermen themselves seemed to change for the better. "To those guys," an NMFS employee had said, "an environmentalist is the lowest thing there is." But now the tunamen were making all sorts of cooperative noises. As if to show just how willing they were to live with the Marine Mammal Protection Act, they even improved on their record in 1979 — killing 18,400 dolphins of a 41,160 quota.

The tunamen were taking more care and working their nets more skillfully, and that was part of the reason why the dolphin kill dropped so dramatically. But it wasn't the whole reason. Usually skipjack tuna averages about one-quarter of the total Pacific tuna catch. But 1978, 1979, and the first half of 1980 have proven banner years for skipjack, with catches running well over double the average and making up close to half of the total tuna take. The important thing about skipjack is that they don't associate with dolphins the way yellowfin do. So the tunamen have been filling up on skipjack, making fewer sets on yellowfin with dolphin, and thus killing fewer dolphins.

Meanwhile, as required by law, the NMFS was again censusing the dolphins to see how well the affected species and stocks were making out. The scientists met in La Jolla, California, in late 1979 to compare their data. Their new population estimates were substantially lower than the 1976 figures, but, they reported, *"this should not be construed to be an indication of large stock decreases between these years."* Instead, *"these differences are mainly the result of better estimation procedures"* (*Report of the Status of Porpoise Stocks Workshop,* 1979, NMFS Southwest Fisheries Center, Administrative Report No. LJ-79-41, pp. 53–54, italics in original). Indeed, with one exception, the scientists felt that the dolphin populations were in good shape. That exception was the northern offshore spotted dolphin, whose numbers had dropped to somewhere between 34 and 55 percent of the optimal level. In NMFS-ese, the stock was "depleted."

NMFS published a new set of rules to cover fishing on dolphin. The 1980 quota for the second half of the year was amended to 16,850, and the 1981 quota was cut to 16,640. A few gear and method changes were ordered, and sundown sets — running the purse seine around dolphins within an hour and a half of sunset — were prohibited because of the greater threat they posed to the dolphins. All this the tunamen could live with. But what set them off was the provision that no take of northern offshore spotted dolphins would be allowed, not even when the animals were trapped accidentally.

That had the tunamen screaming. The northern offshore spotted dolphin is the species most often found with yellowfin. According to the American Tunaboat Association, the ban would cut the yellowfin take in half.

The NMFS wanted to hold hearings on the new rules in February 1980, but the American Tunaboat Association filed suit to push the hearings back to the original April date. The tunamen knew they were in for a fight, and they wanted time to gear up for it. Some of the tunaboat owners began looking for a way out. In March five of the largest American purse-seiners, all of them over 1,000 tons in capacity, transferred their registry to Mexico, along with five Dutch-registered boats from the Netherlands Antilles. The plan — set in motion by Ralston Purina, whose Van Camp

Sea Food Company is a major tuna-boat owner and fish processor in San Diego — was to base the ten vessels in Ensenada and truck part of their catch to the Van Camp cannery in San Diego. There was no doubt why the purse-seiners were moving. August Felando told the press, "We are operating under a sort of doomsday type of regulation on porpoise mortality, and there isn't anything we can do about it."

Hearings on the new rules were held at the Scripps Institution of Oceanography in La Jolla before administrative law judge Hugh Dolan, who took the volumes of conflicting testimony under consideration. Everyone — NMFS, tunamen, environmentalists — waited. In their dark moments, some of the fishermen were no doubt wondering what it would be like working out of a Mexican port instead of an American one.

Then a downturn in local international relations closed that avenue of escape. The United States and Mexico had had a bilateral agreement allowing American vessels to fish in Mexican waters for a $20 annual fee. The Mexicans then refused to renew the treaty and declared that the tuna within their 200-mile limit were permanently resident and thus a Mexican resource. To fish for them, foreign vessels would have to pay a substantial fee. The United States said that Mexico was talking nonsense, the tuna were migratory and thus an international resource. The Mexicans repeated their demands, and when American purse-seiners entered their waters and took tuna, six of them were seized, their nets and catches confiscated, and heavy fines levied. The United States immediately clamped an embargo on tuna imports from Mexico, ending the marketing plans of the transferred Ralston Purina purse-seiners.

The government came to the rescue of the American tunamen. Hugh Dolan gave his final opinion to Richard Frank, who as chief of the National Oceanic and Atmospheric Administration had the power to make the ultimate binding decision on dolphin quotas. Dolan argued that the NMFS's conclusions rested on evidence — basically reports from observers on two research ships and an aerial survey — too shaky to warrant so drastic a curtailing of tuna fishing. Frank agreed. He set a yearly quota of 20,500 dolphins, including 11,890 northern offshore spotteds, for each of the next five years. Frank also told the press this was as low as the quota could go. Further progress, which is to say fewer dead dolphins, appeared unlikely.

Without doubt the tuna-dolphin tangle is a less knotty problem than it was ten years ago. Far fewer dolphins are dying these days than were. But the bottom line remains grisly: dolphins succumb in the nets to supply products of no greater importance than tuna salad and cat food. The old wrong remains; there's just less of it now.

Nor should the government's assurance that most of the affected dolphin species are in good shape be believed without question. Hidden away in all those reports and tables are a couple of disturbing facts. For one thing, the reported number of dolphins killed excludes seriously injured animals. If a dolphin is rescued from the purse seine with a broken beak or flipper but is still alive, the animal is counted as living even though it is unlikely to survive long. The government figures that the number of serious injuries equals about 5 percent of the kill. That 5 percent could be low — after all, "serious injury" is a judgment call — and the true kill might be a good bit bigger than the official statistics let on. For another thing, the dolphins that smother in the nets are disproportionately young animals and lactating females. Removing these animals changes the dolphin population's age and sex makeup. If there are too few young to replace the old as they die off and too few females to keep reproduction going at the needed pace, then the stock could in time decline far more drastically than the simple arithmetical kill totals indicate. Yet NMFS data count only the total population, not its sex or age characteristics. A demographic disaster, such as what befell the common dolphins of the Black Sea, could be in the making, but we won't know about it until it has already happened and the dolphins, the leaping lads before the wind, are gone for good.

To ensure that that sad and silent day never comes, action has to be taken. Several avenues are open.

Who captains the tunaboat and its operations makes a big difference. A random survey of 29 purse-seiners showed that only three boats killed nearly half of the dolphins taken by all of them and that one vessel, a regular Attila the Hun of the eastern Pacific, knocked off 15 percent of the killed dolphins all by itself. Legal sanctions could be imposed to make sure that such destructive captains never leave the dock.

But the law needs more than new sticks. It ought to dangle a few carrots as well. The fishermen need the dolphins as guides to the tuna, and it is in their interest to preserve the animals. That interest, though, is long-term, hard to put in the dollars and cents that worry the captains from day to day. What the law needs is incentives to make saving dolphins attractive right now. Captains and owners who take special care to safeguard the mammals should gain financially, and those who are sloppy should pay. A quota arranged by boat rather than by the whole fleet might help, since the dolphin-killers would fill their quotas faster than the others and have to quit fishing sooner, thus costing them a substantial portion of their catch and reducing the competition for the remaining, more careful vessels. Likewise, it does no good to demand environmental enlightenment from our own fishermen and then to import tuna from other countries whose vessels care nothing about killing dolphins. Evidence that foreign ships perform worse than United States standards should be cause for embargo — a loss that would hurt the offender, since the United States consumes about 20 percent of the world's tuna catch. The law for such embargo exists; it needs thorough enforcement.

The tunamen and many of the NMFS's scientists and technicians see the tuna-dolphin tangle as a gear problem. The key, in their view, is making the purse seine safe for dolphins. To show just what a difference good gear can make, they point to the experimental voyage of the *Elizabeth C.J.*, cosponsored by the NMFS and the Porpoise Rescue Foundation, which is funded by the tuna industry. The purpose of the voyage was to test some gear ideas on a first-class vessel with a first-class crew. The results were impressive. The *Elizabeth C.J.* made 45 sets on dolphin, encircling over 30,000 of them all told, and killed only 16. Kenneth Norris, who was aboard for part of the voyage, felt that eight or even ten of the eleven dolphins he saw killed were victims of experimental manipulations of the net by scientists trying out ideas or hunches and would not have died in normal fishing. If the whole industry did as well as the *Elizabeth C.J.*, then only about 600 dolphins would die in purse seines each year.

But the *Elizabeth C.J.* had a lot more going for it than improved gear. Nicholas Lavolouis, the master, and Manuel and Joe Jorge, the fishing captains, are as good as they come. The vessel also enjoyed near-perfect weather, eliminating many of the problems with the net caused by dirty wind or sea. Even more important, the *Elizabeth C.J.* was fishing during the closed season, without competition from other purse-seiners. That competition grows fiercer each year. Besides the increasing number of American boats, the foreign fleet has doubled in the past decade, putting around 350 tunaboats into the eastern Pacific, all of them scrambling hard for a share of the catch. The pressures on each fishing captain to get as much tuna as he can as quickly as he can — and dolphins take the hindmost — are immense.

The purse seine is a dolphin-killer, pure and simple. Tinker with it as they will, the scientists and technicians will merely shave the kill, not end it. To stop the kill of dolphins altogether — which is, after all, what the Marine Mammal Protection Act demands — some alternative way of taking yellowfin tuna must be found.

The first, and apparently only, alternative comes from Stanley Minasian, who picked the brains of various marine biologists and garnered the experience of tunamen. The idea, simply, is to attract tuna by simulating a dolphin herd. The boat tows an array of large ropes, which by moving across the surface create much the same pressure waves as leaping, swimming dolphins. At the same time, dolphin phonations are played into the water, and small amounts of dolphin feces and sloughed skin, gathered from oceanarium tank cleanings, are released. Tuna have an exquisite sense of taste, and some researchers think that the feces and skin guide tuna to dolphins in the first place. The boat tows its ropes and trails skin and feces until a school of dolphinless tuna arrives to investigate. Then the ropes can be brought in, the purse seine deployed, and the yellowfin sacked up and frozen, without any risk to dolphins.

Of course, the proof of the pudding is at sea, and there Minasian has still been unable to test the idea. To prove incontrovertibly that the system works, Minasian needs strong evidence, which boils down to observation by qualified scientists of good reputation and film footage from every perspective, including aerial. But the price tag for chartering a ship, renting a helicopter, amassing the camera equipment, and feeding everyone for a couple of months runs into the hundreds of thousands of dollars. That kind of money Minasian hasn't been able to raise.

So the idea languishes, and the tangle remains. Not that a few knots haven't been at least loosened. NMFS takes its legal mandate seriously and continues to try to solve the insoluble problem of censusing the dolphins in all those millions of square miles of Pacific Ocean. The tunamen have kept their pride but have also tried hard to adapt to the demands of the law. And the environmentalists have come down from space and given credence to the tunamen's financial needs. Yet dolphins still die and they will keep on dying until somebody comes up with a workable, profitable way of taking yellowfin tuna without endangering dolphins.

If our past history of exploiting cetaceans is any guide to the future, then we can say good-bye now to the Pacific dolphins. But perhaps the progress that has been made is a good omen, boding well for what is to come. Perhaps someday the tunamen will fish and the dolphins will frolic, and the one will not be a danger to the other. It seems such a simple thing to hope for.

Meanings

S I X

It has been said that great art is the night thought of man.
It may emerge without warning from the soundless depths
of the unconscious, just as supernovae may blaze up suddenly in the farther reaches of void space. The critics, like
astronomers, can afterwards triangulate such worlds but not
account for them.
— L O R E N   E I S E L E Y

Legend and Literature

We love our dogs and admire our cats, but for dolphins and porpoises we feel a unique fraternity. People will tell you that this feeling
belongs to our age alone. It doesn't. It goes back to a point before history
and art where religion and mythology blur. Humans and dolphins blur as
well. What looks like the one may well be the other.

The Indians of the Amazon and Orinoco basins in South America believe that during the carnival season river dolphins come ashore disguised
as young men to seduce maidens. The Maoris of New Zealand know why
dolphins accompany boats. A warrior named Ruru cursed a rival with an
oath reserved for chiefs and killed not only the rival but also a dolphin
that happened unluckily to be nearby. To punish him for the blasphemy,
a tribal priest ordered Ruru into the body of the dolphin and told him to
pilot every canoe traveling along the coast.

A truly wonderful story comes from Ha'apai in the Friendly Isles of
Polynesia. In the old days, the men lived only to fight and to boast of their
bravery. They spent their time making weapons, fanning hatred for their
numerous enemies, and planning raids. While the Ha'apais boasted and
planned, the Tongans, fiercest of their enemies, were paddling toward the
island in force. So stealthily did the Tongans come that the Ha'apais had

no warning until the war canoes beached on their doorsteps. Bravely the Ha'apai warriors threw themselves against the invaders, but, caught off guard, they could not gather their forces to counterattack. The Tongans cut the defenders down, smearing the island with blood. The maidens of Ha'apai knew the fate awaiting them. They fled to a high hill and hid, hoping to escape the notice of the blood-lusting Tongans. Six days they remained on the hill without food, and on the seventh hunger drove them down. As the women emerged from cover onto the beach, Tongan warriors spotted them and gave chase. The young women, crazy with hunger and fear, ran into the water and, diving from the reef, leapt into the sea. The Tongans followed, sure of an orgy of easy slaughter. Then the Tongans stopped; terror fixed them in place. The maidens had become dolphins, leaping through the waves. Fearing whatever trick the gods of Ha'apai might work on them next, the Tongans took to their canoes and fled the island. The few surviving Ha'apais emerged from hiding. From that remnant came new generations that remembered the maidens become dolphins and obeyed a strict tabu against killing the creatures.

The Greeks, like the Polynesians, were people of the sea and the islands, and dolphins likewise figure prominently into their mythology. When Apollo crossed to Delphi to found his temple, he took the form of a dolphin. A dolphin saved Telemachus, the son of Odysseus, from drowning, and Odysseus commemorated the event by taking the smiling dolphin as the emblem for his shield. When Amphitrite, whom Poseidon wanted as wife, hid in a sea cave, the god sent a dolphin to fetch her back. In return for the favor, Poseidon set the constellation Delphinus (Dolphin) in the summer sky between Aquila, the eagle, and Pegasus, the flying horse.

A Homeric hymn of the fourth or fifth century B.C. attributes the origin of the dolphins to crime, sacrilege, and mistaken identity. A pirate ship coasting a headland spied there a beautiful young man wearing the royal purple. Already counting the ransom money, the pirates seized the youth, ignorant that they had taken the god Dionysus. They soon found out, though, that their prize was at least a trifle unusual. Ropes fell off the youth as if untied. The helmsman cried out that the young man must be a god and should be freed, but the captain ordered the sail up and the ship to sea. Wind filled the sheet, but the vessel did not move. Wine smelling of ambrosia streamed along the deck. A vine coiled up and about the mast, flowered, and set fruit. The captain, now shaking with fear, told the helmsman to put in to shore, but it was too late. Dionysus had become a lion; roaring, he pounced on the captain. Fearing a like fate, the pirates threw themselves into the sea and were instantly reborn as dolphins. Dionysus spared only the helmsman.

The theme of justice in Dionysus and the pirates figures in another Greek tale. This story concerns Arion, who was a poet and singer of distinction in about 700 B.C. After residing in the court of Periander, king of Corinth, Arion traveled to the Greek colonies of Italy and Sicily to enter music contests and there earn his fame and fortune. He found both in great measure. Intending to return to Corinth, Arion chartered a ship, whose crew, once they learned of the passenger's wealth, resolved to kill him and steal his money. Arion asked to go out singing. Standing on the afterdeck, the poet gave a final performance, then jumped overboard. But Arion did not drown. Attracted by the sweet music, dolphins found the singer in the sea, and one of them bore him on its back to the land. Thence Arion traveled to Corinth, where he told the king of the sailors' treachery. Periander, skeptical of Arion's farfetched story, put the poet under guard and summoned the crew. When asked what had become of Arion, the sailors answered to a man that he had disembarked, hale and hearty, at an Italian port. Out stepped Arion, catching the sailors with the lie in their mouths. To honor Poseidon for the return of his fortune, Arion erected a bronze statue of a dolphin rider at the place where the animal had brought him ashore.

Arion is but one of the many dolphin riders in Greek mythology. The classic among these stories is the tale of the boy of Iassos. The boys of the town, hot and sweaty from running and wrestling in the town gymnasium, ended their exercise with a dip in the sea. A dolphin watching from afar fell in love with the most beautiful of the boys and approached. At first the boy was afraid and fled, but the dolphin's persistence drew him and he came soon to love the creature. The boy and the dolphin were inseparable, playing together as the townspeople watched and envied the dolphin for winning the affection of so beautiful a boy. Often the boy rode on the dolphin's back far, far out to sea, beyond the sight of land and the eyes of the watchers. But always they had to return; the world is too imperfect to support perfect love. One day the boy, exhausted from play, threw himself upon the dolphin's back so hard that the dorsal fin impaled him mortally. The dolphin felt the dead weight atop it and saw the dark blood billowing into the sea. Grief-stricken, the animal ran aground near the gymnasium and died.

Universal and eternal meanings crowd together in this tale. The story is deeply, hauntingly erotic. The love between the dolphin and a boy described as beautiful seems homosexual at first glance. Indeed, in classical art, the dolphin appears conventionally as a phallic image. Likewise, the boy's death on the dolphin's fin has a second, priapic meaning. But this theme is not the sum of the tale. The dolphin is a creature of the sea, the mother of all things. Its name comes from the Greek for womb, *delphis*. The dolphin is thus female as well as male. Its body conceals its own single gender, yet its form conveys the sexuality of both genders. The dolphin of Iassos is total Eros.

The dolphin's love is pure and innocent, and that innocence and purity extends to all aspects of its being. The Cilician poet Oppian writes: ". . . Diviner than the Dolphin is nothing yet created; for indeed they were aforetime men and lived in cities along with mortals, but by the devising of Dionysus they exchanged the land for the sea and put on the form of fishes; but even now the righteous spirit of men in them preserves human thought and human deeds." Dolphins are the good in us; they know not the bad. They befriend us willingly from the goodness of their natures, but we are sometimes so perverse that we betray and kill them. "The hunting of dolphins," Oppian declares, "is immoral, and that man can no more draw nigh the gods as a welcome sacrificer nor touch their altars with clean hands but pollutes those who share the same roof with him, whoso willingly devises destruction for dolphins."

As they succeeded to leadership of the known world, the Christians borrowed and rewrote the dolphin lore of Greece and Rome. Dolphins had served pagan gods like Poseidon; now they became agents of the one true deity. Dolphins saved St. Martian, St. Basil the Younger, and Callistratus from untimely martyrdom, and when the corpse of martyred St. Lucian of Antioch was thrown into the sea for the crabs and sharks, a dolphin bore the body to Drepanum for proper burial. Dolphins decorated baptismal fonts, where the life of the Christian individual began, and they appeared on gravestones, where it ended, to mark the faith of those below. The Christians tamed the dolphin's sexuality; the animal became a sign of marital love and fidelity.

As Christianity turned back on itself to create what we now call the Middle Ages, dolphin mythology, like much classical learning and art, passed into obscurity. Bits, though, survived. Sailors, who often spent weeks or months away from land and fresh meat, relished the flesh of the small cetaceans. Eating it, however, apparently left the sailors feeling guilty. They believed that dolphins could tell from a corpse's smell whether the dead man had ever eaten dolphin meat. If he had abstained, the dolphins would bear the corpse ashore. But if he had indulged, the dolphins returned the favor and devoured his carrion on the spot.

After the Renaissance rekindled interest in classical literature, dolphins returned to art. References to dolphins appear in Spenser's *Amoretti* and *The Faerie Queene*, Shakespeare's *Twelfth Night*, La Fontaine's *Fables*, Keats's "Endymion," Shelley's "Witch of Atlas," Wordsworth's "Ruth," and Browning's "Fifine at the Fair." These dolphins, however, are not so much real dolphins as allusions to classical myths, creatures on the same literary order as satyrs and wood nymphs.

In our own time, the literary tradition of the dolphin as a humanlike creature of play, innocence, and eroticism has reappeared. The dolphins John Lilly describes in *Man and Dolphin* and *The Mind of the Dolphin* draw heavily from the tradition, although Lilly, a frightful writer, appears wholly innocent of literature. Two works published in recent years make explicit use of the literary character of the dolphin.

Leo Szilard's *Voice of the Dolphins* takes the form of a thesis by an anonymous scholar in an undated future and details how the world comes to its senses and rids itself of thermonuclear weapons. The cause of this sudden and unusual rush to public sanity lies in the Vienna Institute, a cooperative research facility staffed by both Russian and American scientists. The scientists bring the dolphins as research subjects, but the dolphins, proving far more brilliant than anyone had imagined, end up telling the scientists how to design their experiments. The dolphin-directed research captures all the Nobels in medicine and physiology in one year. The dolphins turn their attention to politics and begin a series of television programs designed to clarify the real stakes in various international issues. The dolphins' wisdom serves as an antidote to political irrationality, and in time the powers great and small swear off the Bomb.

Szilard was a physicist and an architect of the Bomb. He was also one of the first to see the ultimate stupidity of atomic weapons, and he led the scientists who opposed the use of the atomic bomb against Japan. Szilard lost that round, but he continued the fight against atomic and hydrogen weapons. He put his solution to the arms race in the mouths of his dolphins. No political problem, the dolphins say, is as complex as the scientific problems solved in physics in the first half of the twentieth century. The scientific problems were solved quickly because scientists subjected them to continuous discussion with the purpose of clarifying the truth. Politics, by contrast, seeks only to persuade. If political issues were discussed in the same clarifying manner as scientific problems, they could be solved with the same speed. It takes dolphins to perceive this truth. In their innocent wisdom, unpolluted by the machinations of politics, the dolphins see the true way and guide humans to it.

Robert Merle spins a good thriller in *The Day of the Dolphin*. Two experimental dolphins, named Fa and Bi, learn to speak English through their blowholes. This first breach in the wall between humans and animals is closed quickly by a federal intelligence agency that uses the dolphins to plant a bomb and set the stage for World War III. Only after the harm is done do the dolphins see the evil purpose they have been put to. Sobered but not yet cynical, the dolphins pilot their original trainers on a voyage across open sea to tell the world the truth.

Both Merle's and Szilard's dolphins owe their existence to the dolphins of classical Greece. The creatures are wise innocents, unseduced by human rationalization. They see the truth we will not allow ourselves to see. That truth may destroy them, as it nearly does Fa and Bi. But because the dolphins love us, they guide us selflessly to salvation.

The commonality that humans feel with dolphins is lacking in stories about whales, with one notable exception. This tale comes from Polynesia. The Tahitian version concerns Putu, queen of the island of Nuku Hiva in the Marquesas. Putu was a widow and a woman of great aloofness and royal reserve. When she felt the need to escape the burden of rule, she took to sea on the back of the great bull sperm whale Tokama. Putu had twin daughters, and Tokama twin sons, and the three women traveled the seas astride the three whales. Seen from shore, their hair streaming black in the wind, their bodies sheened with water, the women seemed distant and alluring sirens.

After one such journey, Putu returned to Nuku Hiva to find a captive named Kae among the business awaiting her decision. Kae came from Upolu, 1,500 miles away, and he claimed that he had washed ashore from the wreck of his ship. Putu knew him to be lying; she said he had come to steal a wife. But even Putu did not realize how completely treacherous Kae was. He had in fact come to Nuku Hiva to make good a boast to kidnap and wed the whale-riding twin princesses. Kae claimed, though, that if he was allowed to return to Upolu, he would never come again to Nuku Hiva. Putu's daughters suggested that rather than risk a ship and crew, Tokama should bear Kae home. Putu agreed reluctantly. Kae left on Tokama, the whale's riderless sons taking his flanks.

Kae rode Tokama into the harbor of Upolu and drove the unsuspecting animal onto a shelf of coral. The villagers set upon Tokama with axes and spears, cutting and slashing the stranded beast, skinning him alive and stripping off his blubber. He fought back bravely, crushing many an islander that ventured near his flukes, but finally Kae got a foothold on the whale's head and drove a spear deep into his skull. Thus died the first whale by human hand.

Tokama's mutilated flippers had drifted out to sea, and his sons, each seizing one in their mouths, sped back to Nuku Hiva. Putu already suspected treachery; the priests had read ill omens. The grisly evidence brought by Tokama's sons proved Kae's evil. No longer, Putu knew, would whales serve humans, but she requested one final favor of the sons of Tokama. Her daughters asked that vengeance be theirs, since they had unwittingly sent Tokama to his death. Putu agreed. The twin daughters of Putu set out on the backs of the twin sons of Tokama for the island of Upolu.

Coming ashore in the midst of a fierce storm, the princesses caught Kae, trussed him like a pig for market, returned with their prize to Nuku Hiva. There a priest cursed Kae and with a bamboo knife sacrificed him to the gods. Putu bade the sons of Tokama take to open water, well away from humans and their ships. In sadness, the people of Nuku Hiva lined the beach and watched the twin plumes of the whales' spouts fade and finally disappear in the distance.

This Tahitian story contains many of the same elements as the dolphin myths of Greece: the sexuality of the women on the bull whales, the innocence of the animals, the treachery begotten by human evil on the beasts, even the notion of a time before time when whales and humans lived together.

In Japanese folklore and mythology, the whale is an oversize brute with bluster to match. One such folktale tells a story we know best in Aesop's version of the hare and the tortoise. The whale was bragging: "I am the greatest animal in the sea." The sea slug laughed. The whale angrily challenged the sea slug to a race, and the sea slug agreed to meet the whale in three days. The sea slug gathered all his sea slug friends and asked each one to travel to a beach and there await the whale. The sea slugs tumbled away to their destinations. On the appointed day the whale and the sea slug met and decided to race to the beach at Kohama. The whale surged away, leaving the sea slug tumbling slowly in his powerful wake. At the beach the whale called out, "Sea slug, sea slug, where are you?" And the sea slug who had been waiting called back, "What, whale? Are you only now arriving?" The sea slug suggested that they race next to Shimoda, and off the whale swam. Again the whale called, "Sea slug, sea slug, where are you?" And again the waiting sea slug called back, "What, whale? Are you only now arriving?" The whale and the sea slug raced over and over, always with the same result, and finally the whale admitted his defeat.

The whale's exaggerated sense of self-importance provides the motive in another tale from Japan. Soon after the Daibatsu, a great bronze image of the Buddha, was set in place in Kamakura, news of its immense size reached the whale in his far northern home. The whale dismissed the report; he refused to believe in anything bigger than himself. But the stories kept coming, and the whale had to find out the truth. A shark offered to swim south and measure the statue for the whale. When the shark came

Hourglass dolphin (*Lagenorhynchus cruciger*)

to Kamakura, he enlisted the help of a rat, which scampered about the base of the Daibatsu and counted its steps. The whale, hearing the count of 5,000 rat-paces from the shark, set off to Kamakura himself. More than curiosity drew the whale; his intent was malicious. He put on magic boots and walked ashore to the shrine housing the Daibatsu, but he was too big to squeeze through the door. A priest came out and, with a straightforwardness remarkable for someone talking to a whale in boots, asked the huge animal its business. The whale demanded to know the height of the Daibatsu. At the commotion, the Daibatsu came out and was surprised to meet a creature of its own immense size. The priest measured both statue and whale with his rosary and found the Daibatsu two inches less than the whale in height and girth. Satisfied that he was yet the biggest of the big, the whale went back to the sea and swam home to the north. The Daibatsu returned to its pedestal, from which it never again stirred.

In the Western tradition, the whale of legend, myth, and art is a creature of ambivalent origin and import. The majesty of the animal's size instills wonder and awe, a sense of life on a scale transcending the human and suggesting the divine. But that same size betokens a monster, a seagoing terror ready and waiting to seize and crush the puny men who trust themselves to the waters.

Biblical references to whales call upon both poles of meaning. Whales, though, are and were uncommon in the Mediterranean, and the Biblical writers, like all Hebrews, were confirmed landsmen. They concocted their whales from hearsay and ignorance. The result was a most curious creature. Job (41) describes Leviathan: "Canst thou draw out leviathan with an hook? . . . Who can open the doors of his face? His teeth are terrible round about. His scales are his pride, shut up together as with a close seal. . . . He maketh the deep to boil like a pot; he maketh the sea like a pot of ointment. Upon earth there is not his like, who is made without fear." Leviathan combines characteristics of the crocodile with those of a great whale: scales and fierce teeth, the sea-boil of surfacing. In Isaiah (27:1) Leviathan is most certainly the crocodile: "The Lord with his pure and great strong sword shall punish leviathan the piercing serpent; and he shall slay the dragon that is in the sea." But no crocodile inspired the Psalmist (105:25–26): "So is this great and wide sea, wherein are things creeping innumerable, both small and great beasts. There go the ships; there is that leviathan, whom thou hast made to play therein."

The greatest whale tale in the Bible is, of course, the story of Jonah. God told Jonah to go to Nineveh to preach against that great city's great wickedness. Instead, Jonah headed in the opposite direction and boarded a ship. God sent a storm to stop him. The sailors, realizing that Jonah was the source of their peril, cast him into the sea, and a whale swallowed Jonah. The prophet spent three days and three nights in the belly of the creature, no doubt pondering the high cost of disobedience. When the whale spat Jonah upon the land and God again told him to go to Nineveh, Jonah was on his way. He preached that Nineveh would be destroyed for its sins, and the message struck home. The Ninevites arrayed themselves in sackcloth and ashes, adhered to the strictest fasts, repented of their evil. The conversion pleased the Lord so much that He spared the city the destruction He had planned for it. This turn of divine events, though, angered Jonah, who wanted to see the city in ruins. He prayed to be allowed to die if his enemies lived, but God instead gave Jonah an object lesson in true mercy.

Only a blind and unthinking faith could lead one to take the story of Jonah literally, and that same faith, insisting on the accuracy of trivial detail, misses the point of the tale. The Book of Jonah tells, with terrific directness and power, of a once-parochial tribal God extending his concern to the Gentiles. This is a beautiful and rich story built upon the appropriately beautiful and rich image of the whale of the deep as an instrument of the divine and universal will.

The *Konungs skuggsjá*, a Norwegian work dating from about 1250, resolves the dualism of the whale in Manichean fashion by dividing the species into good and bad. Bad whales like the sperm, the orca, and the narwhal attacked ships unprovoked. Sailors feared them so much that they dared not even mention their names at sea. But good whales like the rorquals helped shipwrecked sailors and drove shoals of fish into the fishermen's nets. The rorquals even enforced civil tranquillity among the fishermen. If they fought and bloodied the water, the rorquals herded the fish away until peace returned.

Many folktales from the medieval period have whales that are not at all what they seem. Unsuspecting sailors seeking refuge anchor alongside what they think is an island and go ashore. The island proves to be a whale, which awakens and sounds, pulling the ship down and drowning all hands. The most famous character to so involve himself with a whale was St. Brendan, the Benedictine who left his native Ireland in 565 to find the Promised Land of the Saints. He put ashore to say Mass on what he thought was an island but was really a whale. Brendan's presence and God's intercession transformed the whale into a real island. The whale-become-land was called St. Brendan's Island, and people gave its existence such credence that explorers searched for it in the Atlantic west of the Canaries as late as the mid-eighteenth century.

Folktales about whales appear to have passed into obscurity around the fifteenth century. Possibly the reason lay in the rise of commercial whaling. The whale changed from a huge, powerful, near-mythological creature to a self-propelled tub of high-income lard. People tell stories about great animals, not about natural resources, and the whale was no longer a fit subject. Yet it remains remarkable that for all the hustle and bustle and the considerable human energy that went into whaling, so little writing came from it, even at its zenith in the nineteenth century. Popular novels by such writers as W. H. G. Kingston, C. L. Newhall, and Alexandre Dumas are few, and books of enduring merit are fewer yet. I count three.

The first is *The Cruise of the Cachalot* by Frank Bullen. Orphaned at nine, Bullen grew up in the streets of London like an urchin straight from the pages of Dickens and went to sea at twelve. In 1875, only eighteen but already experienced at sea, Bullen found himself in New Bedford and signed onto a whaling vessel bound for the Pacific. Bullen later left the sea for a job as a clerk. He wrote about his whaling adventures on the side, and when the book sold well, he gave up his job to write full time. In the seventeen years remaining before his death, Bullen wrote 36 books and uncounted articles, almost exclusively about the sea. The story of the dolphin-maidens of Ha'apai comes from one of his works.

The Cruise of the Cachalot is simply a journalistic account of a sperm-whale voyage in the closing years of Yankee whaling. The virtue of the book is that Bullen writes clearly and entertainingly of the tools, customs, and perils of whaling. Bullen handles people less well, and when he turns to moral questions, particularly the plight and turpitude of the ordinary seaman, he is always cloyingly conventional and sententious.

The second volume is *Marine Mammals of the North-western Coast of North America* by Charles M. Scammon. In the 1850s Scammon was a successful whaling and sealing captain along the Pacific reach from northern Alaska to central Mexico. He later joined the U.S. Revenue Cutter Service, forerunner of the Coast Guard, and while still in the service began writing his observations of the sea. In time Scammon became a roving correspondent for the *Overland Monthly,* a magazine whose editors and principal writers were Bret Harte and Mark Twain. Scammon set to work on *Marine Mammals,* drawn in part from articles published in the monthly, and the book was published, to good reviews and poor sales, in 1874.

Scammon wished to present a catalogue of the marine mammals of the American Pacific and to write a history of the Yankee whaling industry. Scammon's zoological reporting was so accurate that even now scientific authorities cite the book as a standard reference. But the true literary worth of the book appears in Scammon's essays on whaling and most particularly in his description of lagoon whaling in Baja California. Scammon could write, and when he put his pen to the things he knew intimately, he wrote very well indeed. The book provides another instance of an eternal literary irony: that language is often most beautiful when it tells of the worst horrors, the bloodiest butcheries, and the most venal cruelties.

The third volume about whaling is the best. It is in fact one of the best of all books ever. It is *Moby Dick*.

Herman Melville was born August 1, 1819, in New York City, the third child of a brood that finally totaled eight. Melville's father, Allan Melvill (the family adopted the -*e* after he died), imported dry goods and notions from France, a trade that provided an ample living. Despite the tie to commerce, the Melvilles laid claim to a modestly noteworthy heritage. Allan Melvill traced his ancestry to Scotch-Irish settlers who came to New England before the Revolution and even farther back to a thirteenth-century Scots nobleman. Melvill's wife, Maria, came from old New York Dutch stock and was the daughter of the Revolutionary hero Peter Gansevoort, who commanded the defense of Fort Stanwix against Burgoyne.

Pretensions to aristocracy bought little in the depression that ground down the American economy in the mid-1820s. Allan Melvill's import business faltered. He borrowed from his father and from his brother-in-law Peter Gansevoort, but bankruptcy swallowed him up anyway. Ruined, Allan Melvill had to move his family to Albany, where they depended wholly on Peter Gansevoort for support. The strain of misfortune proved too much for Melvill. He passed into a delirium and died early in 1832.

The death forced Herman and his older brother, Gansevoort, to work. Gansevoort opened a fur and cap store and Herman clerked first in a bank and later in his brother's store. But business was not the Melville forte. A financial panic in 1837 finished Gansevoort's store, and Herman ended up teaching school near Pittsfield, Massachusetts. He studied surveying in hope of getting a job on the Erie Canal, but nothing came of the plan. Gansevoort, meanwhile, had gone to New York to read law and make his fortune. He turned up a job for Herman as a cabinboy aboard the *St. Lawrence,* a packet on the run between New York and Liverpool.

Not yet twenty, Herman Melville had no idea of the rigors of life at sea: the cold, the constant wetting, the seasickness, the vile food. But worst of all was the casual brutality misnamed discipline and the sadistic pleasure with which the officers broke the men to their will. Life improved little when the *St. Lawrence* docked in Liverpool. The waterfront was a gauntlet of brothels, grog shops, criminals, and landsharks. As soon as the ship returned to New York, Melville headed inland and found another position as country schoolteacher.

When the school year ended, Melville and a friend went to Illinois in search of a piece of the new, western action, but they found nothing. By fall, Melville was back in New York looking for a job, any job. Months passed; he found nothing. Desperate, prompted as always by a strong sense of family responsibility, Melville made up his mind. He traveled to New Bedford and signed on with the sperm-whaler *Acushnet.* On January 3, 1841, the vessel set sail.

Valentine Pease, the captain of the *Acushnet,* was a brutal man who ended up in an asylum on Martha's Vineyard, but he proved a capable enough whaling master. Only two months into the voyage, the *Acushnet* offloaded 150 barrels of oil at Rio de Janeiro, a good take in so short a time. The vessel rounded the Horn and plied the Onshore and Offshore grounds, richer by far than the waters of the Atlantic, but the *Acushnet*'s early run of luck had been its last. Whales were sighted infrequently, killed rarely. The days were long, repeated monotonies broken only by occasional visits from other whalers — "gams" as the whalemen called them.

During a gam, Melville met a young man named Chase, son of Owen Chase, first mate during the well-known *Essex* disaster. Melville questioned young Chase about his father's misadventure, and as he was about to leave the ship for his own, young Chase gave him a copy of his father's published narrative of the wreck. Melville devoured the book.

In November 1820 the *Essex* had been whaling close to the area where the *Acushnet* was then cruising, about midway between the Galápagos and the Marquesas. While the whaleboats were chasing a pod of sperm whales, a large bull whale doubled back and rammed the ship twice. Badly stove, the *Essex* settled in a matter of minutes. At first the 21 men in their three whaleboats lay close to the sunken hulk in the hope that another whaler would come along, but when none appeared, the officers decided to run to land. Fearing hurricanes to the north and cannibals to the south, they struck out for the coast of South America, 3,000 miles away. Eight survived. The journey was a three-month ordeal in a chamber of maritime horrors that included cannibalism. The story of the misfortune captured the essence of man's fate at the hands of nature, a meaning clear even to the elder Chase, who was a man of action, not philosophy. He wrote: "I . . . endeavored to realize by what unaccountable destiny or design . . . this sudden and most deadly attack had been made upon us by an animal never before suspected of premeditated violence." The story came home strongly to Melville. In his copy of the narrative he scribbled: "The reading of this wondrous story upon the landless sea, and so close to the very latitude of the shipwreck, had a surprising effect on me."

At the moment, though, Melville's immediate concern was more monotony than man's fate. He had already had his fill of whaling, yet the empty hold meant that the voyage might last years. For months Pease had kept the vessel out of the sight of land, but finally he relented and headed toward Nuku Hiva to pick up supplies and to give the men a needed break among the brown Polynesian women. The ship anchored, the orgy commenced. Melville supposedly remained aloof from the debauch, but he liked what he saw of the island well enough to stay. He and another crewman, Richard Tobias Greene, called Toby by the men, resolved to jump ship, choosing to desert just before the *Acushnet* was due to set sail in the hope that Pease would leave rather than stay around to hunt them. They guessed right. The *Acushnet* left Nuku Hiva two hands short.

Melville and Greene wanted to stay away from the bay of Nuku Hiva lest the newly established French authorities decide to prove their sovereignty by imprisoning them for desertion. The two traveled across the mountainous spine of the island to reach the Happars, a tribe with a reputation for friendliness, but they blundered instead into the valley of the Typees, held to be the bloodlustiest of cannibals. The Typees, though, proved hospitable and generous. But they could do nothing for a crippling infection in Melville's leg, and Toby hiked over to the bay to fetch medicine. He never returned. Melville remained another three weeks, and then, perhaps bored with paradise or fearful that he would end up as the entrée at some feast — the fate he feared had befallen Toby — he signed onto the *Lucy Ann*, an Australian whaler.

Any ship that took on a crewman like Melville, too hurt to work, was in a bad way. The *Lucy Ann* was in the worst way. The captain was incompetent and sick. The mate, an able seaman when sober, was never sober after breakfast. Many of the crew had deserted, and several more men had been clapped into irons and put off on a warship for fostering mutiny. It was all the ship could do to sail, much less lower for whales. When the captain's condition worsened and the mate made for Tahiti and medical aid, he lay outside the harbor to keep the remaining crew from deserting. The crew in turn filed grievances with the British consul, who without considering the complaints sided with the captain and the mate against the men. Open mutiny threatened, and several troublesome sailors, including Melville, were arrested and put ashore to face trial. Nothing happened. The episode proved an embarrassment to the consul, and the prisoners simply walked away from the broken-down jail. Melville and his companion, John Troy, a surgeon who had gone to sea after suffering the nineteenth-century equivalent of losing his license, hoboed about as they pleased. Eventually Melville tired of the footloose life and he put himself back in harness on the *Charles & Henry* for a six months' whaling voyage toward South America and up to Hawaii.

The *Charles & Henry* fared no better at whaling than the *Acushnet* had. Melville disembarked in Honolulu and collected the little money due him. There he took a job as a merchant's bookkeeper until the end of summer. He joined the United States navy and two days later shipped out on the frigate *United States*. The ship took Melville back to Tahiti and Nuku Hiva and then on to Callao and Valparaiso, with a side trip to Mazatlán. Then the ship headed south, doubled Cape Horn, and worked up the Atlantic. On October 3, 1844, the *United States* dropped anchor in Boston. Herman Melville had come home.

Melville's mother and sisters loved the stories he told of the South Seas — likely heavily edited in some respects and embellished in others — and their adulation combined with the family's new-found financial security prompted the young man to write about his experiences. Apparently Melville had thought little about writing as a career before this time. Except for two typically adolescent essays published in a local newspaper before his voyage on the *St. Lawrence*, Melville had written nothing, not even a notebook or a journal. He had been thoroughly and painfully purposeless before he sat down to write. There he discovered something addictive, and from that day on Melville had the writer's monkey on his back.

Once Melville decided to write, he went at it with energy and will. He set to work in the spring of 1845 on his adventures among the Typees. In July, Gansevoort Melville, who was traveling to a new job in London, took part of the manuscript with him to seek an English publisher. Herman Melville himself arranged for publication in the United States. Early in 1846 the American and English editions of *Typee* appeared.

Critics dealt favorably with the book, but some doubted its authenticity, intimating that *Typee* was fiction fobbed off as fact. Melville's veracity was proved by the appearance of the one person who could corroborate the story: Toby. While seeking medicine for Melville's leg, Greene had been tricked aboard a whaler that left Nuku Hiva with him unwillingly aboard, and he, as ignorant of Melville's fate as Melville had been of his, was living in Buffalo. He wrote to a local paper and vouched for Melville's honesty.

Melville had in fact embellished the true story of his stay among the Typees, lengthening his visit from three weeks to four months and adding dramatic effect by playing up his fear of the cannibal feast. But those details are only literary devices and not the point of *Typee* and its picture of an idyllic Polynesia. Modern readers are likely to miss that point, to dismiss the book as just another South Seas romance, without realizing that *Typee* set the style for such romances. Before *Typee* the only books about Polynesia were a missionary's account of taking religion to the pagans and the story of a surgeon shipwrecked on Tonga. Our own fantasy vision of the green island in the blue sea and its brown-skinned, clean-limbed people originated with Melville. The South Seas stories of Jack London, Robert Louis Stevenson, and James Michener owe their nature to Herman Melville.

Typee reveals just how counter to the spirit of his times Melville's art would take him. America of the 1840s was convulsing with religious fervor, and Yankee missionaries breathing hellfire were fanning out over the globe to peddle the gospel to the pagans. Melville had lived among those pagans; their world was happy, secure, and kind. He had also lived among Christians, in the cities, in the forecastles, in the brothels, and the grog shops. He doubted the Christian claim to moral superiority. "The term 'savage,' " he wrote, "is . . . often misapplied, and indeed when I consider the vices, cruelties, and enormities of every kind that spring up in the tainted atmosphere of a fevered civilization, I am inclined to think that so far as the relative wickedness of the parties is concerned, four or five Marquesan Islanders sent to the United States as missionaries, might be quite as useful as an equal number of Americans dispatched to the Islands in a similar capacity."

Meanwhile Melville was writing about his experiences on Tahiti. *Omoo*, which means "rover," appeared in the spring of 1847 and was as well received as *Typee* had been. Suddenly Herman Melville, who only three years before had been a maintopman in the navy, was a full-maned literary lion.

Flush with success, Melville married Elizabeth Shaw, a long-time friend of his sister Helen and the daughter of Judge Lemuel Shaw, Chief Justice of the Massachusetts Supreme Court. But Melville had more reputation than money. He could keep his family eating only if he kept himself writing. Established in a house in New York, Melville composed *Mardi,* which started out as another Polynesian romance but soon changed course. Melville's increasing concern with questions of fate and the universe twisted the tale into a symbolical and often ham-handed allegory. The reviewers derided it. Sales were poor, and Melville, only recently a father, went back to his desk to try to write himself into the respectable prosperity he wanted and felt he deserved.

Melville considered his next two books, *Redburn* — based loosely on his apprentice voyage to Liverpool — and *White-Jacket* — drawn from his days in the navy — to be potboilers. In a letter to Lemuel Shaw he described the novels as "two *jobs,* which I have done for money — being forced to it, as other men are to sawing wood." Actually, the books are honest tales well told. They did not prove the financial boon Melville wished them to be. The mediocre sales fell short of the advances Harper Brothers had made. By 1850 Melville owed Harper over $700, and the company refused him any more cash.

That summer, Melville took his family on a vacation to the Berkshire Hills around Pittsfield, Massachusetts, an area that was earning a reputation as a retreat for artists. There, at a picnic on Monument Mountain, Melville met Nathaniel Hawthorne. Melville deeply admired Hawthorne's work, and he wished to establish a literary friendship to rival Goethe's and Schiller's. The two men and their families spent considerable time together, and both moved to the Berkshires permanently, there to think of weighty matters and to write well and truly.

Something big was brewing in Melville. Before leaving New York, he had begun a book on sperm whaling. Once settled into his farm in the Berkshires, surrounded and supported by a family of wife and son and mother and unmarried sisters, Melville launched into the best writing of his career. It was not a wholly happy time. Melville's letters talk of blank days, his head aswim with images and his hand too halt to set them down on paper. Financial worries upset his artistic serenity. "The calm, the coolness, the silent grass-growing mood in which a man ought to compose," wrote Melville to Hawthorne, "that, I fear, can seldom be mine. Dollars damn me. . . ." But when the work went well, it went very well indeed.

He was deeply into his story of Ahab's pursuit of the white whale when he stopped working to plant his farm in the spring. The chores done, Melville picked up where he had left off and finished the book. At first he called the novel *The Whale,* the title borne by the English edition. But at the last moment before publication in 1851, Melville changed the Harper Brothers' version to *Moby Dick.*

Melville knew he had written a big book, and he pinned on it a great many hopes, both financial and literary. The critics, however, disliked *Moby Dick.* They wanted a sea tale, not a metaphysical descent to the roots of the universe, and they set upon Melville not so much for the book he wrote as for not writing the one they wanted him to.

Despite, or perhaps because of, his disappointment at the reception of *Moby Dick,* Melville went back to work. *Pierre,* published only thirteen months after *Moby Dick,* is an equally ambitious work, a psychological novel with intimations of incest that were more than a little unseemly in that thoroughly repressed age. *Pierre* drew heavy critical fire and destroyed the last of Melville's literary reputation. As if to symbolize his fall from literary grace, a fire in the Harper warehouse consumed the plates of Melville's novels and most of the copies of his books in stock.

The next years of Melville's life were difficult. The strain of writing two massive books in such a short time to such poor response sapped Melville's health and depressed his spirit. Money remained a problem. Melville's friends, including Hawthorne, tried to find him a post as a consul, but they came up empty. The writer went back to writing. He did magazine sketches and stories, and in 1855 he published *Israel Potter,* a cleverly satirical tale about an undercover hero of the American Revolution. The following year a collection of Melville's stories called the *Piazza Tales* came out, and in 1857 *The Confidence Man* appeared, the last of Melville's novels to be published in his lifetime. Melville tried to cash in on the lecture craze of the late 1850s, but even though he set his fee low, he was little in demand.

Melville's family hoped that a long sea voyage might restore his health and well-being, and his brother Thomas, captain of the clipper ship *Meteor,* invited him along on a trip to San Francisco. The family hoped as well that the sea air might clear Melville's head of literary delusions and bring him back ready and able to buckle down to something with a profit in it. Of late, Melville had been experimenting with verse. Elizabeth Melville wrote confidentially to her mother: "Herman has taken to writing poetry. You need not tell anyone, for you know how such things get around."

The voyage did little to help Melville's health, but when he returned and found that a batch of poems left with his brother Allan had been rejected by all the publishers, Melville went along with the family's plans to find him a consular appointment under the newly elected Lincoln. While Melville was considering whether to take a post in Glasgow or wait for something better, Lemuel Shaw died suddenly. Although Shaw's executors deducted the judge's many generosities to the Melvilles from Elizabeth's inheritance, enough remained for them to move from Pittsfield to New York and settle into a comfortable way of life.

The Civil War broke out. Melville hated war as the collective criminality it is, but he honored true heroism and bravery. He wrote poem after poem about the war, and in 1866 a collection called *Battle-Pieces* was published. The book's seeming patriotism helped Melville gain appointment as a deputy customs inspector in the port of New York. At the age of 47, Herman Melville took his first regular job.

There is something pathetic in the image of an artist of Melville's caliber checking manifests and crawling around cargo holds in search of contraband. The author, though, enjoyed the work. The job took him out of the house, brought in regular money, and gave him enough time to keep up his writing. Melville's health improved, his emotional and mental outlook brightened. He had made it through the long dark night of the 1850s, and he was willing to keep on going. In a letter he wrote: "I once, like other spoonies, cherished a loose sort of notion that I did not care to live very long. But I will frankly own that I have no serious, no insuperable objections to a respectable longevity. I don't like the idea of being left out night after night in a cold churchyard."

Melville remained with the customs office for almost twenty years, and most of his writing in that period was devoted to *Clarel*, a long and obscure poem about an intellectual odyssey through the Holy Land. *Clarel* never would have seen the light of day had not Melville's uncle Gansevoort subsidized publication.

Late in 1885, Melville returned to full-time writing. A number of legacies had been left him and Elizabeth, and with a comfortable income guaranteed Melville resigned the customs post and gave himself over to his real work. He warmed up by assembling and publishing two collections, *John Marr and Other Sailors* and *Timoleon*. Then Melville turned his art to a story he had heard in the navy long years before, a story about a mutiny, a drumhead court, and a triple hanging aboard the warship *Somers*. Melville shaped this bare tale into an excellent novel about good, evil, law, and social necessity. He called it *Billy Budd*. He put the manuscript away, apparently intending to work further on the story.

He never got the chance. A heart ailment put him in bed and worsened daily. On September 28, 1891, Herman Melville died.

The event aroused minor interest, not so much from sadness at Melville's passing as from surprise that he had lived so long. The New York *Daily Tribune* gave only one paragraph to Melville's obituary, which declared *Typee* the best of his books. The *Tribune,* however, did much better than the *Critic,* a literary journal, whose editors were so uncertain of Melville's past that for a death notice they plagiarized a few hundred words from a compendium on American literature.

Melville's death did direct some attention anew to his work, but only to the early novels. Melville had foreseen his fate decades earlier. "Think of it!" he wrote to Hawthorne as his reputation slid. "To go down to posterity is bad enough; but to go down as a 'man who lived among the cannibals!' " Melville saw *Typee* being handed to toddlers along with their gingerbread. The writer who felt the darkness pressing in all around had become an exotic entertainment for children.

In England, curiously, interest in Melville's later work, and particularly in *Moby Dick,* remained alive in certain literary circles. But in the United States, interest in Melville came not from recognition by fellow artists but from the tenacity of a hungry graduate student. Graduate students are forever looking for original dissertation topics, and such a search brought Raymond Weaver to Melville, who was so obscure he had to be original. Melville's descendants gave Weaver access to the family records, among them the trunk in which Elizabeth Melville had packed all her husband's papers, including the unknown *Billy Budd* manuscript. Weaver wrote an adulatory critical biography of Melville in 1921 and oversaw publication of *Billy Budd* three years later. The Melville boom was on. Critics and literary historians by the score wrote essays and books and monographs about Melville's life, about Melville's work, about Melville's psychiatric state, about Melville's quarrel with God, about Melville's layers of language, and so forth. The scholarly fuss has elevated Melville to the topmost rank of American writers, where he belonged all along.

There is a moral to this journey from forecastle to fame to obscurity to posthumous greatness, a moral that Melville himself, with his keen sense of ambiguity, could best capture. If he is watching — and surely so resilient a spirit has survived these many nights in that cold churchyard — he must be chuckling in his long gray beard.

Literary greatness entails a paradox. A book becomes a classic because it is great reading, but because it is a classic, people are afraid to read it.

The paradoxical fate of *Moby Dick* has been worsened by all the voluminous critical writing about the book, which, while supposedly seeking to make the work clear, has with its ponderous academic theorizing succeeded only in making it more obscure.

The story line of *Moby Dick* is simple enough. A young man, who gives only his adopted name Ishmael, has sickened of life. He decides on a bold step. "With a philosophical flourish Cato throws himself on his sword," says Ishmael. "I quietly take to ship." Ishmael has made a whaler his ship of choice. He is drawn by a curiosity in "such a portentous and mysterious monster" as the whale and by a desire to see "the wild and distant seas where he rolled his island bulk."

Ishmael arrives in New Bedford late on a Saturday, after the last packet for Nantucket has sailed. He must wait till Monday. Ishmael settles into a tavern, where, owing to a lack of space, he has to share a bed with a pagan harpooner named Queequeg, a tattooed Maori who wears a top hat and carries his long gleaming iron wherever he goes. Ishmael and Queequeg become the dearest of friends, and they make a pact to ship out together from Nantucket.

The vessel they choose is the *Pequod,* "a cannibal of a craft," with teeth and jaw ivory of the myriad sperm whales killed by the ship in its many voyages. The *Pequod*'s owners are a curious Quaker pair, Bildad and Peleg, who tell Ishmael of the even more curious one-legged captain, Ahab. On the quay Ishmael and Queequeg meet Elijah. Declaiming like a prophet, Elijah warns of a doom awaiting the *Pequod* and her crew. Ishmael dismisses Elijah as an idiot ranting nonsense. He and Queequeg join the assembling crew, and on a cold Christmas day, the *Pequod* "blindly plunged like fate into the lone Atlantic."

The *Pequod* strikes south and east, and its bounded world afloat reveals itself layer by layer. The officers are three: Starbuck, the prudent Quaker and reliable company man; Stubb, a loudmouth, but brave as his arm is strong and his lance sharp; and Flask, a dullard so absolutely dull that he thinks everyone and everything as dull as himself. The harpooners, too, are three, pagans all: Queequeg, whom we know; Tashtego, a Gay Head Indian from Martha's Vineyard, descendant of true warrior hunters; and Daggoo, an African so huge that a "white man standing before him seemed a white flag come to beg truce of a fortress." The crew comprises loners and misfits like Ishmael. He calls them isolatoes.

In the early days of the journey, the crew has it only on faith that Ahab is aboard. The captain stays in his cabin, leaving the routine running of the vessel to Starbuck. Not until the ship reaches warmer latitudes does Ahab appear on the quarterdeck. The sight of the captain sends a surge through the crew. Ahab is tall, spare, weathered, superhuman, marked with a livid scar like a brand running down the side of his face. A post of gleaming jaw ivory replaces his leg. He reveals the real purpose of the voyage: to hunt and slay Moby Dick, the same brute that mutilated him. Ahab's emotion, manner, and purpose are dark and unholy, but the passion of his vengeful quest rouses the crew. They all join willingly in his compact of death — all, that is, but Starbuck, the prudent bourgeois and the God-fearing Quaker.

At the first lowering for whales, the crew learns the thoroughness of Ahab's plan for revenge. He has secreted in his cabin a boat crew of pagans headed by a Parsee fire-worshipper named Fedallah, who looks and acts like the very valet of the devil. For an unholy pursuit, Ahab has chosen unholy companions.

The cast of the drama extends beyond the human characters, and as the *Pequod* makes her way through the seas in search of the geyser spout and snowhill hump of Moby Dick, a discourse on whales and whaling paces the story. In exquisite detail, Melville tells of the whale, inventing his own bookman's taxonomy and dividing the cetaceans by size into folio, quarto, and octavo. He describes the anatomy of the sperm whale, the mechanics of the animal's poetic movement, its social organization. He tells, too, the history of whale hunting and the methods, tools, and customs of the industry.

Ahab's thirst for vengeance swells from singlemindedness to grand obsession. He abjures all comfort, all social contact. He casts his pipe into the sea. He will not stop to gam with other whalers. He merely passes alongside and calls out, "Hast seen the white whale?" Finally his question brings a positive reply. He boards the other whaler and meets a captain who mirrors his own mutilation and repair: Captain Boomer has a jaw ivory arm, a memento of Moby Dick. Twice since, Boomer has crossed wakes with the white whale, but he will never pursue him again. Boomer is brave and unafraid; he is also sensible. " 'He's welcome to the arm he has,' " Boomer explains to Ahab, " 'since I can't help it, and didn't know him then, but not to another one. No more White Whales for me; I've lowered for him once, and that has satisfied me.' " Ahab, angered by Boomer's resignation, quits his ship in such haste that he shatters his ivory leg.

Now, like the first puff of breeze rustling a slack sail, the novel stirs toward its climax. Starbuck, a Christian caught among pagans by circumstance, tries to dissuade Ahab from his vengeance, but Ahab refuses and dares the mate to mutiny. Starbuck, always a man of duty, knows himself bound to that duty and thus to Ahab's fate. A fever fells Queequeg, and, thinking himself about to die, he orders a coffin from the carpenter. But Queequeg recovers, and the coffin replaces a lost and worthless lifebuoy. Ahab has the smith forge a special harpoon edged with his own razors, and he tempers the iron meant for Moby Dick in blood drawn from the pagan harpooners. In a typhoon, the eerie glow of St. Elmo's fire lights up the masts, a sign to whalemen that death is near. Ahab defies the omen. He grasps the lines to the lightning rods and lets the forces of the sky pulse in him. The storm reverses the ship's compass; Ahab fashions a new one. His course is set; there is no turning aside. The *Pequod* comes across the *Rachel*, which lowered for Moby Dick on the day before and is searching for a lost boat carrying the captain's son. That grief-stricken master pleads for help in the search, but Ahab, deaf to all but news of Moby Dick's proximity, pushes on. The last thread connecting Ahab to the human community has been severed. He is utterly and violently alone on the open sea, his crew but tools to his murderous purpose.

The chase begins. Ahab himself, hoisted into the rigging, spots Moby Dick's snowhill hump and sings out. At the lowering, Moby Dick sounds. The boats wait on the still surface, until the white monster bursts up from the depths and crushes Ahab's boat in his corkscrew jaw. The whale hunts for foundering Ahab in the wreckage, but it is driven off when the *Pequod* sails down upon it. All the men and the blood-tempered iron are picked up.

The second day of the chase begins, again with the masthead watches peeling their eyes for the white leviathan. Moby Dick appears in a breach, casting his whole great bulk into the sky. The boats are lowered; all make fast. Moby Dick tangles the lines, draws Stubb's and Flask's boats together, smashes them to kindling. Then Moby Dick turns on Ahab's boat and breaks it apart with his battering-ram head, scattering everyone and everything into the foaming sea and its confusion of half-drowned sailors and sinking gear. The *Pequod* picks them up, all save one: Fedallah is missing.

The sun dawns on the third day of the chase. Hours pass, all is still, then the white whale, as persistent as Ahab, appears. The one remaining boat is lowered, accompanied by a column of sharks that bite at the dipping oars. The whale sounds, appears like a mountain rising from the sea. Moby Dick's sides and back bristle with harpoons, and snared in the tangle of their lines lies the corpse of Fedallah. A shudder passes through Ahab; Fedallah has prophesied that he would go on before. But Ahab presses the chase, drives nearly onto the whale, and sinks his iron into the beast's flank. The line parts. Moby Dick turns away and rushes the *Pequod* itself. All aboard watch in terror as the monster smashes into the

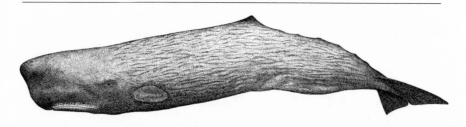

Sperm whale (*Physeter macrocephalus*)

bow at full speed, opening a hole so great that the harpooners in the mast-heads can hear the water flooding in like a cascading torrent. The sight inflames Ahab. No longer is he merely vengeful; now he is vengeance. He darts an iron into Moby Dick. The singing line fouls, loops itself about Ahab's neck, and jerks the master into the sea and death. The *Pequod* sinks, its whirlpool sucking all in.

One man survives: Ishmael, saved by happenstance on the buoy of Queequeg's coffin and rescued by the "devious-cruising Rachel, that in her retracing search after her missing children, only found another orphan."

It is obvious to any reader of *Moby Dick* that the tale proper suggests deep and disturbing thoughts. It abounds with symbols, grand language, ruminations, pronouncements, and portents. What, the critics ask themselves, is the real meaning of *Moby Dick?*

The answer depends on the critics. For Van Wyck Brooks, Moby Dick is much like Grendel in *Beowulf,* and the book, like that ancient poem, springs from the Nordic consciousness of hard and endless struggle against the elements. Percy Boynton gives the novel a political cast by making the white whale a symbol of the vested privilege that lames the spirit of creative man. Newton Arwin holds that Ahab meets his doom because he is guilty of the Christian sin of pride or the Greek flaw of hubris. Henry Murray applies the paradigm of Freudian psychoanalysis to *Moby Dick.* Ahab is the id, Moby Dick the ego, and good steady Starbuck the super-ego. Murray goes further. Ahab's death by a line attached to the whale shows Melville's umbilical fixation to the imago of the mother and signals his submission to parental conscience. His animus against civilization springs from thwarted Eros, symbolized by Ahab's maiming, which indicates amputation not of the leg but of the penis. I am curious to know what meaning Murray gives Ahab's peg-leg.

Murray's interpretation of *Moby Dick* and others like it assume that the novel is a consistent allegory in which fictional events and characters stand for events and characters in real life. But the assumption of allegory doesn't square with *Moby Dick* itself.

It is apparent that Melville had no intent to write an allegorical novel. About the time he started the story, he wrote to Richard Bentley, the English publisher of *White Jacket,* and described the work in progress as "a romance of adventure based on certain wild legends of the Southern Sperm Whale Fisheries, and illustrated by the author's personal experiences of two years or more, as a harpooner." Nothing indicates that Melville's plans changed greatly while working. In a letter to Mrs. Hawthorne, Melville responded to a now lost letter from her husband about *Moby Dick:* "I had some vague idea while writing . . . , that the whole book was susceptible of an allegoric construction, & also that *parts* of it were —but the speciality [sic] of many of the particular subordinate allegories were first revealed to me after reading Mr. Hawthorne's letter. . . ." Melville clearly had no symbolical plan worked out in his head when he sat down to write, a plan implied in critical interpretations like Murray's. Instead, he worked with his meanings and symbols as they revealed themselves in the course of the writing. Criticism assumes deductive art; Melville wrote *Moby Dick* inductively.

There is nothing surprising in this realization. It would be startling to find that Melville had reasoned himself into some metaphysical position and then constructed a novel in which he explicated the logic of his speculation. Had Melville begun this way, he would have written an essay or a tract, not a novel. *Moby Dick* sprang from artist's fire, not from some philosopher's perverse desire to couch his message in the obscurest form possible.

Writing is as much discovery as explication. The writer, hot and disturbed, sets to work, trying to put names and rhythms to whatever is burning inside him. Often, as he writes, the fire leaps up without warning or it flares out suddenly or it sets other fires that cause new pain, offer new light, and cast new shadows. The words come out, events form themselves, the novel takes on shape, all to the surprise of the writer, who is the site of the fiery confusion but by no means the author of it. And when he finishes the book — or, more likely, abandons it when the coals are cold — the novel remains as the oracular testament of his private inferno.

Moby Dick itself evidences Melville's shifting fires. The tone of the novel changes distinctly once the *Pequod* leaves Nantucket. To that point, the book reads like some rollicking tale of adventure and camaraderie on the order of *Typee* and *Omoo.* But at sea the mood becomes as dark and slow as a Shakespearean tragedy. The story's viewpoint changes as well. The first 22 chapters are told in the first-person voice of Ishmael. Thereafter, Ishmael appears as both character and narrator, but the book describes incidents, such as the confrontation between Ahab and Starbuck and conversations of Stubb and Flask and of Ahab and Fedallah, that Ishmael could not have witnessed or heard. Melville never resolves the problem of viewpoint in this book; he lets the story have its head, trammeling as it will the conventions of the novel. Nor did Melville get all the facts in the story battened down. The reader meets a sailor named Bulkington, who earns a whole chapter and seems bound for some important role in the novel, then all but disappears. Stubb appears as second mate, drops to third, then reappears as second. The helmsman of the *Pequod* stands his watch now at a jawbone tiller, now at a wheel. The loose ends show a writer concerned more with the rush of his tale and the blaze of his fire than with the fine points of literary puzzlemaking. Besides, Melville had deadlines to meet, money to earn. "This whole book," he writes in *Moby Dick*, "is but a draught — nay, but the draught of a draught. Oh, Time, Strength, Cash, and Patience!"

Moby Dick is no allegory, yet the novel does indeed bristle with allusions and symbols. Melville used Biblical names and references by the fistful, beginning with the famous opening line: "Call me Ishmael." The Biblical Ishmael was the son Abraham fathered by the slave woman Hagar when Sarah, his wife, was barren. When Sarah produced her own son, she prevailed upon Abraham to drive out Hagar and her son to prevent their claim on the inheritance. Ishmael became the unwanted wanderer, fated to roam. Ahab was a king of Israel who "did more to provoke the Lord God of Israel to anger than all the kings of Israel that were before him" (I Kings 16:33). Ahab worshipped the false gods of his evil and heathen wife, Jezebel. He met a divinely willed end in a battle he entered even though told that he would be killed there. The thorn in Ahab's side throughout his ungodly reign was the fiery and fearless prophet Elijah, whose name Melville gave to the wild man Ishmael and Queequeg meet on the quay in Nantucket. Obviously Melville chose these names very much on purpose. The Biblical references fit the characters of *Moby Dick* and tell us something more about each of them.

Objects appear and reappear throughout *Moby Dick* in way suggesting special significance. Consider the hammers. Queequeg brandishes a tomahawk. Ahab, tortured by his unforgiving and relentless pursuit, asks the blacksmith to take the heaviest maul to his master's pained head and smooth his ribbed brows. As the *Pequod* sinks, Tashtego on the masthead holds a hammer high. Birds, too, become portents and signs. Ishmael describes the wonder of his first sighting of an albatross. Birds accompany

Ahab's boat on its way to battle Moby Dick. A sky hawk pinned between Tashtego's hammer and the mast goes down with the *Pequod,* and over the gulf of the ship's sinking fly wheeling crowds of sea fowl. Then there are the ropes. Ahab is hoisted into the rigging on a rope to look for Moby Dick. A guard watches the secured end on deck, lest some sailor mistakenly untie the wrong line and drop the captain fatally to the deck. Roped to Ishmael, Queequeg goes over the side to handle the blubber hook. If the harpooner should fall into the sea, Ishmael will surely follow him to doom. He says: "Queequeg was my own inseparable twin brother; nor could I any way get rid of the dangerous liabilities which the hempen bond entailed. . . . I saw that this situation of mine was the precise situation of every mortal that breathes. . . ." Without the whaleline the whaleman could not earn his living, but with it he often meets his death. Flying coils snare limbs, maiming the unfortunate horribly or jerking him to graveless death in the sea. "The graceful repose of the line, as it silently serpentines about the oarsmen before brought into actual play — this is the thing which carries more of true terror than any other aspect of this dangerous affair. But why say more? All men live enveloped in whale-lines. All are born with halters round their necks; but it is only when caught in the swift, sudden turn of death, that mortals realize the silent, subtle, ever-present perils of life. . . ."

Each of these signs or symbols or portents is first and foremost the object itself. The hammers, birds, and lines are all real and true. The whales of *Moby Dick* share that same reality. It seems that many of the critics who tease arcane meanings from *Moby Dick* have never seen a whale, much less been close to one on open water in a small boat. They know nothing of the humbling immensity of the event, and when they feel that sense in Melville's writing, they attribute it to willful symbolism. The meaning comes instead from the truth of the writing.

Melville took great pains to be accurate. He wrote from experience, but he based the novel on more than his own observations. The White Whale came from tales Melville likely heard in the forecastle, tales collected in an 1839 *Knickerbocker Magazine* story "Mocha Dick" by J. N. Reynolds. Melville hunted up every reference to whales he could — his collected quotations appear at the opening of *Moby Dick* as "Extracts (Supplied by a Sub-Sub-Librarian)" — read all the historical material on the whaling industry, and kept Thomas Beale's *Sperm Whale Fishery* close by as a handy and authoritative reference as he composed. Never afraid to debunk silly or fantastic ideas about whales, Melville set down a basic description of the sperm whale that cetology has improved little.

But Melville doesn't stop with scientific objectivity. As with his hammers, birds, and lines — all of them drawn with a keenly observant eye — Melville looks for the meaning in the thing. The whale is the grandest creature in creation: "If ever the world is to be again flooded, like the Netherlands, to kill off its rats, then the eternal whale will still survive, and rearing upon the topmost crest of the equatorial flood, spout his frothed defiance to the skies." Melville makes the whale the embodiment of nature's characteristics, the personification of its various goods and varied evils. Beauty and brutality entwine: two opposites encompassed by one reality. In "The Grand Armada," perhaps the most marvelous chapter in *Moby Dick,* a whaleboat enters a lake of calm within a huge school of sperm whales. The whales come snuffling up to the boats like domestic dogs, and in the waters beneath the gentle creatures hang the giant forms of the mother whales nursing their huge infants. Into this tranquillity breaks a thoughtless horror: a harpooned whale that has parted its line and carries tangled in the trailing ropes a sharp-edged cutting spade. In agony from its wounds, the whale "was now churning through the water, violently flailing about with his flexible tail, and tossing the keen spade about him, wounding and murdering his own comrades."

Whaledom is the essence of nature, and the essence of whaledom is Moby Dick. His power and size provide both magnificence and malevolence. He, too, is both beautiful and brutal. Ahab, though, perceives only the evil that snatched away his leg. "Ahab had cherished a wild vindictiveness against the whale, all the more fell for that in his frantic morbidness he at last came to identify with him, not only all his bodily woes, but all his intellectual and spiritual exasperations. The White Whale swam before him as the monomaniac incarnation of all those malicious agencies which some deep men feel eating in them, till they are left living on with half a heart and half a lung. . . . All evil, to crazy Ahab, were visibly personified, and made practically assailable in Moby Dick. He piled upon the white whale's hump the sum of all the general rage and hate felt by his whole race from Adam on; and then, as if his chest had been a mortar, he burst his hot heart's shell upon it."

Two authors have come closest to capturing the feel of *Moby Dick.* Neither is a critic; both are writers. They show the empathy one artist can feel for the work of another, the very element no formal theory of criticism can include.

The first is D. H. Lawrence, who wrote on Melville in his *Studies in Classic American Literature.* One wonderful aspect of *Moby Dick* is that you do not so much read about the voyage as experience for yourself the fated daily progress of the *Pequod.* Lawrence's essay has a similar character. You feel as if you are looking over Lawrence's shoulder as he reads and listening to him musing aloud. But even Lawrence cannot refrain from putting a single meaning to the White Whale: "He is the deepest blood-being of the white race; he is our deepest blood nature. And he is hunted, hunted, hunted by the maniacal fanaticism of our white mental consciousness. . . . Hot-blooded sea-born Moby Dick. Hunted by monomaniacs of the idea."

In *Aspects of the Novel,* E. M. Forster shows why Lawrence, even though he tries finally to force *Moby Dick* into the pattern of his own work, finds Melville's wavelength. Both Lawrence and Melville are what Forster calls prophetic novelists. Forster does not mean prophecy in the commonplace sense of foretelling future events; he means it in an ancient, Old Testament sense. The theme the prophet-writer chooses is "the universe, or something universal, but he is not necessarily going to 'say' anything about the universe; he proposes to sing, and the strangeness of song arising in the halls of diction is bound to give us a shock. . . . The novel through which bardic influence has passed often has a wrecked air, like a drawing-room after an earthquake or a children's party. . . . Prophecy — in our sense — is a tone of voice." The prophet-writer reaches back; his characters and his incidents sound deep, disturbing chords, often prompting what Melville himself called "the shock of recognition." The experience is by no means intellectual. It happens at some underlying and wordless level where fright and faith compound. Translating the objects in the book, like the hammers or the ropes or Moby Dick himself, into symbols of unvarying meaning savages the story, for naming steals the rhythm of Melville's song. "The essential in *Moby Dick,* its prophetic song, flows athwart the action and the surface morality like an undercurrent," Forster writes. "It lies outside words. Even at the end, when the ship has gone down with the bird of heaven pinned to its mast, and the empty coffin, bouncing up from the vortex, has carried Ishmael back to the world — even then we cannot catch the words of the song. There has been stress, with intervals: but no explicable solution. . . ."

Forster has set down the truth of *Moby Dick.* The book is rough, unfinished, ill-fitting. Its language is sometimes too grand, its point of view uncertain, its story too simple and obvious to move one so. But it does move one so, and therein lies the mystery and the art. Melville reaches deep, rings primal notes that we know but cannot name. He sings of "a blackness and sadness so transcending our own that they are undistinguishable from glory."

In that black, sad sea swims the white island bulk of Moby Dick trailing the haltered corpse of Ahab. That is horror. But there is beauty in this sea, a beauty that sweetens the sadness and brightens the glory. The whale broadcasts it; Ishmael bears witness. "As in the ordinary floating posture of the Leviathan the flukes lie considerably below the level of his back, they are then completely out of sight beneath the surface; but when he is about to plunge into the deeps, his entire flukes with at least thirty feet of his body are tossed erect in the air, and so remain vibrating for a moment, till they downwards shoot out of view. . . . This peaking of the whale's flukes is perhaps the grandest sight to be seen in all animated nature. Out of the bottomless profundities the gigantic tail seems spasmodically snatching at the highest heaven. So in dreams, have I seen majestic Satan thrusting forth his tormented colossal claw from the flame Baltic of Hell. But in gazing at such scenes, it is all in all what mood you are in; if in the Dantean, the devils will occur to you; if in that of Isaiah, the archangels. Standing at the mast-head of my ship during a sunrise that crimsoned sky and sea, I once saw a large herd of whales in the east, all heading towards the sun, and for a moment vibrating in concert with peaked flukes. As it seemed to me at the time, such a grand embodiment of adoration of the gods was never beheld, even in Persia, the home of the fire worshippers. As Ptolemy Philopater testified of the African elephant, I then testified of the whale, pronouncing him the most devout of all beings."

Thus sails Ishmael, the eternal masthead watch and poetic everyman, one eye to heaven and the other to the sea and its whales.

•   •   •   •   •

*I guess we're all, or most of us, wards of that nineteenth
century science which denied existence to anything it could
not measure or explain. The things we couldn't explain
went right on but surely not with our blessing. We did not
see what we couldn't explain, and meanwhile a great part of
the world was abandoned to children, insane people, fools,
and mystics, who were more interested in what is than in
why it is. So many old and lovely things are stored in the
world's attic, because we don't want them around us and
we don't dare throw them out.*
— J O H N   S T E I N B E C K

The Desert Whale

A few begin the ritual about the middle of December, abandoning the metropolitan madness of Christmastime for the silence of the shore. Not that the beach is exactly hospitable that time of the year. A northwest wind blows cold and sharp-edged out of the Gulf of Alaska, and storms have a way of coming up fast and hitting hard. But the early few grit out the weather and watch the sea and its long winter swells. They are waiting for something worth waiting for. They are waiting for the gray whales.

The first ones are single animals, sometimes spaced long hours and even days apart. Then, as December ages, more whales appear, often in small groups. By the beginning of January the whales are passing in daily dozens. As the whales crowd together, so do the watchers. All it takes is a clear weekend in January and the coast bristles with people crowding the headlands and cliffs, eager to see whales.

There is something marvelous in all migratory passages, and there is something particularly impressive about the migration of the gray whales. A couple of years ago, some friends and I hit a perfect day near the lighthouse at Point Reyes north of San Francisco. A violent storm had passed through two days before and left behind it air that was glass-hard and

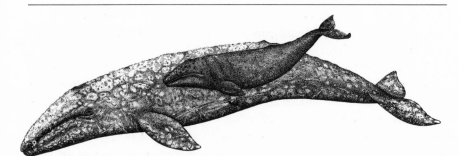

Gray whale and calf (*Eschrichtius robustus*)

bright. From the cliff we could see north for mile after mile past Point Tomales to Bodega Head and beyond. The water from here to there was a plumed processional of spouting whales. They traveled in ranks three to six abreast, breathing in a close-order rhythm. The whales came past slowly, spread from the breaker line to a point perhaps two miles offshore. I tried counting for a while; I had this notion that if I put numbers to the event, I somehow had it in hand. But the whale ranks kept changing course and crossing back and forth across each other, and I soon lost count. I put away my notebook, and the event continued in the absence of mathematics.

Watching the people was nearly as much fun as watching the whales. Some came and saw what they wanted and left; others hung around for hours, entranced by the passing animals. Of all the people the kids were the best. They were excited even before they got out of the cars, and by the time they ran down to the overlook, they were nearly convulsing with expectation. A spout rose. A little blond guy with hair in his eyes screamed, "There's one! I see one!" Then another spout rose, and another and another and another, and the kid kept screaming, "I see one! And there's another one! Do you see it? Do you?" He was jumping up and down and pointing and shouting as hard as he could, and his cheeks flushed below his bangs. There were so many spouts and so many whales that the kid quickly wore himself out, fell silent, sat down with hands and head on knees, and stared. The adults usually just watched; but they shouted and cheered too when a whale, apparently feeding on fish about a mile north, flung himself from the water again and again, a glistening missile rising out of the sea and falling back to his element in a mountainous splash.

Such scenes, repeated again and again along the California coast, make the gray whale the most watched great whale in American waters. Actually this is somewhat curious, in that the gray is neither widespread nor numerous. Its popularity results from a chance mix of cetacean evolution and human demography.

Paleontologists and cetologists consider the gray whale the most primitive of the baleen whales. The animal is essentially similar to an early form from which the right whales and rorquals evolved. As evidence, the scientists point to the gray's small head, short baleen, and few throat grooves. There is also the matter of habitat. The other mysticetes spend part or all their lives in deep water away from land. The gray, no doubt like the large cetaceans of thirty million years ago, is a creature of coastal shallows. But make no mistake about the meaning of *primitive,* an unfortunate choice of words. The gray whale is like the whales of antiquity, but it is hardly obsolete. It has survived because it works. The gray whale fits its world.

The gray's annual cycle of life begins in the Arctic reaches of the Bering, Wrangel, and Chukchi seas. The whales feed almost continuously, largely on a bottom-dwelling amphipod crustacean (*Ampelisca macrocephala*) they capture by sifting floor sediments, a method of foraging unique to the species. By the end of summer the whales are fat. As autumn closes the Arctic ice, the whales turn south.

The object of their journey is Baja California. The trip from Alaska to Mexico covers 3,800 miles, not the 6,000 often reported, but the migration remains the longest undertaken by any mammal. The whales come through the Aleutians, apparently strike out across the Gulf of Alaska toward Vancouver Island, then follow the continent south and return close inshore along California. Once in Mexico, the whales divide themselves up among several wintering areas. Perhaps half go to Bahía Sebastián Vizcaíno and the complex of Laguna Ojo de Liebre (Scammon's Lagoon) opening up from it. Most of the remainder go to either Laguna San Ignacio or to Bahía Magdalena and its environs. A few double Cabo San Lucas and end up along the eastern side of the Golfo de California.

Evolutionary happenstance keeps the gray whale close to shore. For millennia they have come this way, and for almost all that considerable length of time the western coast of North America was only thinly populated. But in this very century, in the latest of the successive conquests of California, people by the millions have moved to the coast and taken up life alongside the whales' migration route. The restricted range of the gray includes the doorsteps of a tenth of the population of the United States. All the coastal Californians have to do to see whales is to step outside and look.

Still, to many first-timers, the sight is disappointing. Everybody has been telling them how big whales are: thirty feet, forty feet, even fifty feet, yards and yards of whale. But sighting from the land conveys little bigness because of the distances involved. People forget they are seeing whales that are as much as two miles away.

The only way to get a real sense of the whales is to see them from the water. Northern California offers ocean conditions by and large too nasty for good whale watching. Monterey Bay and southern California are better. But the best way to see whales well is to take a week and go by boat from San Diego to the desert bays and lagoons of Baja California.

The *Finalista 100* leaves the dock at H&M Landing at 11 o'clock at night, but the trip does not begin then. No boat trip begins with the casting off. That is a preparation, a mere cutting of old ties. The trip properly begins only when the boat passes the breakwater from the harbor into the sea and the first ocean swell rolls the deck under your feet.

Laguna San Ignacio, the whale-watching destination, lies some 450 miles south of San Diego, almost exactly halfway down the Baja peninsula. The vessel runs largely at night, putting the passengers ashore at Todos Santos Island the first day and San Benito del Oeste the second, to see desert plants and birds and seals on the beaches. Whales and dolphins appear regularly at sea, but the true drama begins the third morning out, about two or three hours after dawn. Morning in Baja California means a land breeze, and it blows into the swells, standing the waves up and sparking the air with spindrift. Ahead of the bow breakers roll in ranks, seemingly all the way to the foot of the hazy mountain that actually lies well beyond the deepest reaches of the lagoon. The boat comes in closer and closer to the uncertain line of sea and land, and the breakers grow bigger and more violent. The water shallows fast, greening as the sand bottom comes up. The man at the helm holds the wheel tight and concentrates on what he is doing. He takes the boat in easy. When the *Finalista 100* passes over the bar into the lagoon, only two or three feet of water lie under the keel.

But for all the tension of the moment, only that man at the helm is paying attention. The passengers are crowded against the rails, fingers and arms pointing and cameras ready. The sea all around seems alive with gray whales, dozens and dozens of them. The great backs roll and shine in the sunlight, and at each spout passengers yell and cheer and point until their voices are hoarse and their arms tired but they keep on shouting and pointing.

Deeper into the lagoon the water calms somewhat, and here the boat anchors. The scene is a perfect paradox. A ship rides at anchor in the midst of whales in a lagoon enclosed by flat, treeless, scrub desert. The place is at once real and unreal.

But there is more to watching whales than seeing them from a big boat like the *Finalista 100*. The craft itself acts as a barrier, a kind of final protection between observer and observed. To see whales as they are, you have to go among them on their terms. That means going in an Avon, a small rubber skiff with an outboard. It twists and bends with the chop and splashes water in your face when it runs fast into the wind. And when a whale comes close, you can actually feel the slight disturbance of its wake through the skin of the boat.

Getting close is usually a simple matter. The helmsman guides the skiff out to a channel the whales favor, idles the engine, and lets the boat drift with the tide and the wind. Slowly the whales come past, single whales and cows with calves. Their progress suggests the rotation of a surfacing and submerging wheel, as if the animals more roll than swim. The head comes up and the water spills off, then the blowholes clear and the spout bursts up. The back breaks free of the water as the head goes down. Usually the back slips under then, and only a slick remains, but now and again the flukes kick out, hanging in the air for an instant, and disappear in a propelling wave.

The experience is not, as some would have it, something like bumping into a wise extraterrestrial off a UFO — the technological society's equivalent of Moses' meeting the burning bush. But it is impressive in a way that little else is. The proper comparison is hard to find. The best I know comes from a man who had once hunted hard. He had been out in an Avon that morning, and now it was late in the evening. "You know what it was like," he said to me, and he continued without pause. "It was like one morning when I was on safari. It must have been twenty years ago. I was in the bush early just after dawn, no rifle. I saw this elephant and walked up to it. It was a big bull, all tusks and ears. He knew I was there. We just stood there looking at each other not ten yards apart. Maybe ten minutes passed, maybe half an hour, I don't know. But finally it just ended and he walked off his way and I went mine." He sipped his drink. "Seeing the whales was like seeing that elephant, looking it in the eye and feeling no fear."

The gray as most-watched whale is but the latest chapter in the animal's chronicle. Indeed, the gray whale's history is a microcosm of the changing fortune and fate of all the great whales.

In historical times, there were three populations of gray whales: the California stock, a small North Atlantic population, and an Asian stock that summered in the Sea of Okhotsk and wintered in the Sea of Japan. Aboriginal whalers like the Eskimos and Chukchis hunted grays, at least opportunistically. Archaeological evidence shows that the Indians of Baja California, who often camped by the rich feeding grounds of the lagoons, salvaged drift and stranded carcasses, but they did not actively hunt whales.

The real hunting action began, as elsewhere, with the advent of commercial whaling. The North Atlantic population of grays disappeared by the mid-1700s. The size of the original stock is so obscure and the evidence from the period so colloquial and fragmentary that it is hard to say whether whalemen exterminated the grays or whether they simply finished off the remains of a fast-disappearing population. Not that the Yankees or the Europeans cared much one way or the other. The gray whale was very small potatoes, what with all the sperms and rights to be had.

When the Yankees entered the North Pacific in force in the nineteenth century, they largely ignored the gray whale. A few were killed, but usually only when hunts for other species had come to naught. Even in the whaling boom of the 1820s and 1830s the gray was little sought. It yielded less than half the oil of a bowhead or a right, and its short, coarse baleen had almost no market value. But, with whaling as with all extractive businesses, success inevitably breeds demise, and a sow's ear has a way of looking more and more like a silk purse.

By the mid-1840s, even as the whaling industry peaked, the desirable sperm and right whales were fast growing scarce. There was good money to be had killing whales, but there were fewer whales to kill. Bahía Magdalena was a popular wintering haven for Pacific whalers, and in 1846 the whalemen tried their hand at killing the grays gathered there.

The enterprise lasted only three winters, partly because the price for whale oil fell and made the hunt unprofitable, partly because the gray whale proved unexpectedly ferocious. Unlike other whales, which usually reacted to the harpoon only after the line was snugged against the loggerhead, the gray felt the first bite of the iron and lashed out its flukes in a sidewise blow that smashed men and boat. In the murky narrow channels of the mangrove swamps skirting the bay, the whale had the upper hand, and often the pursued animal would turn unseen and ram the boat before the boatheader even knew the whale was charging. Although all the classic yarns of whaling imitate *Moby Dick* and cast the sperm whale as villain, the gray was far and away the most dangerous to hunt. Japanese whalemen knew the gray's ferocity. They called the animal *koku-kujira,* the devilfish.

The gray whale's reprieve was short-lived, however. The whale oil market recovered and the sperms and rights continued to decline. Once again the whalemen went after the gray, but this time they had something that put the odds more in their favor: the bomb-lance shoulder gun, which allowed the whalemen to finish off a harpooned whale quickly and at a safe distance. The whalemen fast exhausted the grays of Bahía Magdalena.

Enter Charles Melville Scammon. Originally from Maine, where he went to sea young and soon rose to become captain of a coasting schooner, Scammon brought a ship full of would-be gold hunters to California during the Gold Rush. He liked San Francisco well enough to settle permanently, and he joined in establishing the whale and seal oil business in that port. In 1856 Scammon made his first winter whaling trip to Bahía Magdalena and there learned the rewards and dangers of gray-whaling. Before Scammon took his first whale, grays destroyed two boats and damaged the others repeatedly, injuring six of the eighteen officers and men. The survivors patched the boats, splinted the broken bones, and hunted the remainder of the season without mishap.

The whalemen knew there were more gray whales than the few thousand wintering in and about Bahía Magdalena, but they were uncertain where the animals went. Stories from cattlemen, traders, and Indians told of a great lagoon to the north, off the bight of Bahía Sebastián Vizcaíno, but the storytellers were landsmen who knew nothing of the approach from the sea. Scammon was not about to look for that entrance without good cause. The fishhook coast of the bay is a lee shore full of shoals, and no seaman in his right mind would go there unless he had to.

Finally Scammon had to. He had been out all summer and well into the fall of 1857 whaling and sealing with miserable luck. Scammon induced his crew to remain longer at sea in hope of filling out the cargo, and he resolved to find the lagoon he had heard about. He had the tender schooner *Marin* sent down from San Francisco. The *Boston*, Scammon's bark, and the *Marin* met in Catalina, then a working harbor instead of a playground, and headed south. Keeping the *Boston* in safely deep water, he sent the *Marin* ahead to scout. Two days later a boat from the little ship came with news of the lagoon and of a passage from the sea, treacherous but navigable. But when the *Boston* entered, the wind failed, and the bark and its crew spent a sleepless night in the pounding surf. Next day, after an anxious morning of fitful wind, the vessel got through.

A run of bad luck marred Scammon's seeming success. A mishap drowned the ship's carpenter, then a gale blasted the lagoon for three days. When the wind fell, Scammon went deeper into the lagoon, where the whales, almost exclusively cows and calves, congregated in great number.

The boats were lowered, the whales set upon. In the ensuing battle, all the boats were damaged or sunk, and half of the crew was put out of commission with bruises, internal injuries, and broken bones. The men lost their taste for whaling. The next time Scammon sent a boat after a whale, all the oarsmen jumped overboard the moment the boatsteerer threw the iron. "Our situation," Scammon wrote in *Marine Mammals of the Northwestern Coast of North America*, "was both singular and trying. The vessel lay in perfect security in smooth water; and the objects of pursuit, which had been so anxiously sought, were now in countless numbers about us. It was readily to be seen that it was impossible to capture the whales in the usual manner with our present company, and no others could be obtained before the season would be over."

Circumstance brought Scammon to the lagoon; now circumstance forced him into the lethal ingenuity at which the Yankees excelled. He hit upon the idea of ambushing whales with shoulder guns. Whalemen had largely used shoulder guns to finish off harpooned whales. Scammon wanted the shoulder gun and its exploding bomb-lance to do the harpoon's task of securing the whale. Boats were stationed in a narrow channel, and when a whale came by at close range, the marksman blasted a bomb-lance into the whale's viscera, sending a second after the first if he had time. Rarely did the whale die outright, but the wounds generally proved mortal, and in the confined space of the lagoon finding the drift carcass posed little problem. Scammon's gambit worked perfectly. Not another man or boat was lost, and the tryworks were kept busy around the clock. Indeed, the operation proved a whaleman's paradise. Whaling usually was a monotonous search; here the whales were right at hand. Whenever the supply of blubber ran low and another whale was needed, a boat crew rowed out into the lagoon, blasted a whale, attended its agony, and towed it ashore. Scammon took oil in such abundance that he had every available vessel filled, including the bread casks, mincing tubs, and even the trypots, for the trip back to San Francisco. The *Boston* and the *Marin* came through the Golden Gate so heavily laden that their scuppers were awash.

The voyage turned a remarkable profit. Scammon had filled his ship in only eight months at a time when the task usually required four years. His expenses, though, were only a fraction of the long voyage's overhead, and much more of the roughly $15,000 income from the cargo was left for the owners and the crew. Up in the Sierra men were taking gold from the ground; Scammon was pulling it out of the Mexican lagoons.

The next winter Scammon returned, at the head of a horde. He brought a bark and two schooners, and behind him came several dozen hopeful whalers, nine of which actually entered the lagoon. The retreat of the gray whale became a slaughterhouse. By Scammon's estimate, twenty-five whaleboats a day, from late December on into March, pursued the whales, which had no place to run and could only stand and fight. Some of the whalemen hit upon the barbaric method of harpooning the calf, towing it fast ashore, then shooting bomb-lances into the cow as she came after the wounded, struggling young. From sea to shoal, from shore to shore, the lagoon was a harbor of killing, an oddly surreal mixture of men in boats killing Leviathan among sand dunes and cactus. Scammon set it down: "The scene of slaughter was exceedingly picturesque and unusually exciting, especially on a calm morning, when the mirage would transform not only the boats and their crews into fantastic imagery, but the whales, as they sent forth their towering spouts of aqueous vapor, frequently tinted with blood, would appear greatly distorted. At one time, the upper sections of the boats, with their crews, would be seen gliding over the molten-looking surface of the water, with a portion of the colossal form of the whale appearing for an instant, like a spectre, in the advance; or both boats and whales would assume ever-changing forms, while the report of the bomb-guns would sound like a sudden discharge of musketry; but one can not fully realize, unless he be an eyewitness, the intense and boisterous excitement of the reckless pursuit, by a large fleet of boats from different ships, engaged in a morning's whaling foray. Numbers of them will be fast to whales at the same time, and the stricken animals, in their efforts to escape, can be seen darting in every direction through the water, or breaching headlong clear of its surface, coming down with a splash that sends columns of foam in every direction, and with a rattling report that can be heard beyond the surrounding shores. . . . In fact, the whole spectacle is beyond description, for it is one continually changing battle-scene."

The following year Scammon did not return to his lagoon. He may have already sensed that the slaughter, both in the lagoon and by shore stations and coastal whalers prowling from San Diego to Cabo San Lucas, was taking so heavy a toll that catches must decline. Instead, Scammon went farther south to San Ignacio, spotted only two years earlier by his brother-in-law, Jared F. Poole. In the 1860–1861 season, Scammon did try his lagoon again, but the take was so poor that he had to go back to San Ignacio to fill out the cargo. Scammon never again returned to the lagoon that came to bear his name.

The concerted hunt for the gray whale lasted only ten years. Prior to 1854, only about 700 grays had been killed by Yankees, mostly in Bahía Magdalena. Then from 1855 to 1865, the decade of the gray whale, more than 5,000, excluding orphaned calves, were killed in the lagoons, by ships alongshore, on the Arctic summer grounds, and by shore stations in California. Thereafter, the kill fell off, with grays taken largely by happenstance from the shore stations. Even this paltry kill declined until the shore stations went out of business.

Killing in the lagoons had amounted to a selective holocaust that added greatly to the final toll of the dead. Most of the whales in the lagoons were cows, either pregnant, newly delivered, or in estrus. Bulls and juveniles entered from time to time, but they spent much of the winter in the breakers or out in the bay where they were hard to kill and secure. The whalemen concentrated on the easier pickings inside, killing cows out of all proportion to their numbers and destroying their calves with them. The annual crop of young needed to replace the year's dead was cut off short, in effect depleting the population twice, and only a fraction of the available bulls could find mates.

Present estimates, based on the sophisticated statistics biologists use to cloak their guesswork, put the original population of California gray whales in the neighborhood of 15,000 to 20,000. Yankee whaling from the 1840s to 1874 took about 8,000 animals. The population dropped not to the 7,000 to 12,000 straight arithmetic gives, but to somewhere between 2,000 and 5,000 — the precise figure depending on which biostatistical guess you chose to believe. By the end of the nineteenth century and of Yankee whaling, only about one-fourth to one-third of the gray whales survived.

Here arises a mystery, or perhaps a lesson. Four thousand gray whales is considerably fewer than 15,000, but it remains a goodly number of good-sized animals. Yet the gray whale was rarely sighted. Scammon himself wondered "whether this mammal will not be numbered among the extinct species of the Pacific." About the turn of the century, the conventional scientific opinion was that the gray whale had indeed ceased to be. Roy Chapman Andrews wrote, "For over twenty years [preceding 1910] the species has been lost to science and naturalists believe it to be extinct."

The whales were there, the naturalists were missing them, and you begin to wonder why. For one thing, the whales were probably avoiding humans. The gray learns fast who its friends and enemies are, and it doesn't forget. Possibly, the grays, having learned what humans intended to do to them, avoided people and migrated farther offshore than they had been wont. No evidence supports this argument, though.

Observation involves more than the observed; there is also the matter of the observer. The observer-naturalists Andrews mentions had little skill. They watched whales by waiting at the slipway of a shore station to see what the boats towed in. When they saw no dead gray whales, they assumed that the species itself must be also dead. The truth of the matter was that the observers weren't observing.

Economic self-interest makes for finer vision. It took whalers to find the grays again. Floating factory ships had too much draft to enter the lagoons, but they could and did anchor in sheltered harbors along the Baja California coast and send catchers after whales alongshore. Two Norwegian factories operated in the late 1920s, and in the 1930s two American factories hunted the same waters. Of course, the Americans were way behind the times. They were using floating factories when the pelagic factory with its stern slipway had become the basic instrument of high-seas whaling. The Russians used such a vessel in the Bering Sea, and the Japanese took grays in a pelagic operation in the Chukchi Sea. These pelagic expeditions continued even though the 1937 International Agreement for the Regulation of Whaling recognized the gray whale's depleted and endangered population by protecting the species. Neither Japan nor Russia signed that agreement, and they killed as they wished.

When the whaling nations met to draft the 1946 International Convention for the Regulation of Whaling, cetologists again believed the gray whale nearly extinct. A Japanese shore station at Ulsan, Korea, had killed off the Asian stock of grays, and the California population was thought to number only a few hundred. Given the gray's obviously sorry state, the convention protected the species from all hunting except by aborigines and for scientific reasons. The whalers abided by the agreement, not from the goodness of their hearts, but simply because they had no reason not to. The gray whale was commercially nonexistent.

In the years that followed, the gray whale became a swimming contradiction. On the one hand, it enjoyed protection. That meant, so the documents read, that the species was not to be hunted. On the other hand, the gray whale was hunted. That was what the fine print in those same documents allowed.

One of the loopholes was the aboriginal exemption. Alaskan Eskimos killed only about two grays a year, but about thirty a year were taken by Siberian Chukchis. This killing probably had to be; the Arctic natives depended on the whale for food. Then the Russians worked something of a scam. In 1964 the wording of the aboriginal exemption was changed to allow not only killing by the aborigines but also killing by an IWC nation in behalf of its aborigines. The Soviets dispatched a catcher boat to Siberia and increased the kill from the average of 30 a year to 175. The Chukchis aren't eating that much more whale meat. According to Ronn Storro-Patterson, who has been trying to puzzle out the truth through a fog of long distance and Soviet bureaucracy, the whale meat is fed to minks, martens, and other furbearers whose sale to the Russian state fur monopoly brings pin money to the Chukchis and considerable foreign exchange from overseas marketing to the men in the Kremlin.

The other loophole was the scientific permit. In the 1950s, as the few conservationists in the IWC fought and lost to the whalers, it was apparent that science knew very little about whales. Oddly, the gray whale, although the most accessible of all the great whales, was very poorly known. Apart from the writings of Scammon and Andrews, little scientific literature on the animal existed. Dale W. Rice and Allen A. Wolman, then with the Bureau of Commercial Fisheries in Seattle, set up a definitive study of the gray whale. Their report, published by the American Society of Mammalogists with the title *Life History and Ecology of the Gray Whale (Eschrichtius robustus)*, appeared in 1971. The method of the study was classic zoology. Between 1959 and 1969, 316 gray whales — 77 of which were pregnant females — were killed by catcher boats of the Del Monte and Golden Gate Fishing Companies of Richmond, California. The researchers dissected and measured the animals at the shore stations and collected tissue samples such as earplugs and baleen plates for later analysis. Field observations from shore, ship, and plane were made, mostly to count the whales, but the real heart of the study was the cadavers and the data they gave.

The result is most bizarre. The study claims to be of the gray whale, yet there is, in all 142 pages, but one photograph of an intact animal, and that is a near-term fetus cut prematurely from its killed mother. As for the adults, you have to conjure your own image from the charts, graphs, and illustrations of seasonal variations in testis size and other anatomical arcana. The book celebrates parts, not whales. It claims to speak of life from the study of the dead and of ecology from an animal removed violently from its environment. Fiction demands no willing suspension of disbelief as bold as this.

Rice and Wolman's study served an explicit goal: "The California gray whale stock has increased so much that a resumption of commercial exploitation has been considered. As the dearth of basic data on the biology of the species would handicap an effort at rational regulation of the harvest, the Bureau of Commercial Fisheries . . . initiated a research program. . . ." Note that the authors do not question whether whaling should resume; they wish only that it be conducted "rationally."

The study worked to the immediate as well as future benefit of whaling. For their help in the killing of whales for study, the two whaling companies received the carcasses for processing and sale. This proved no small boon. All but six of the whales were killed between 1964 and 1969; that is 310 whales. According to International Whaling Statistics the total American catch in these same years was 1,366. The grays from the scientific sample equaled more than one-fifth of the American commercial catch in those years. Toward the end of the period, the scientific kill of grays provided a substantial portion of the whales killed commercially. The whole American catch of 1968–1969 was 183 whales, of which 74 were grays taken under the scientific permit.

By rights, the Rice and Wolman study should have condemned the gray whale. By the late 1960s, it showed, the population of grays had clearly recovered to about 11,000. That was enough to permit whaling, particularly in a world where whalers were fast running out of whales. Given the example of history, the gray whale's protected days were numbered.

Politics, of all things, saved the animal for a time. As the environmentalism of the 1960s emerged, the whales became prime examples of the irrational destruction called rational exploitation. Popular sentiment opposing whaling grew. Meanwhile, the United States delegation to the International Whaling Commission was moving against whaling. Obviously, it would have looked bad for the United States to politick against the Japanese and the Russians and then apply for a removal of protection from the gray for the benefit of its own vestigial whaling industry. Instead, the federal government outlawed domestic whaling. The gray was among the whale species listed as endangered under the terms of the Endangered Species Act in 1970, and the government canceled the one remaining whaling license the following year. Mexico's strong stand against whaling in the IWC and its 200-mile limit closed Baja California to alien whalers. This foreclosed everything except the Siberian hunt, and the Russians kept on taking grays for the Chukchi fur farms.

The New Management Procedure, under which the IWC labors these days, provides that when an exploited whale stock drops below its optimal level, whaling ceases and the stock is protected. It also provides that if the population recovers, protection is dropped and whaling can begin again. So, at its 1978 meeting, the IWC, following the recommendation of the Scientific Committee, dropped the California gray whale's protection, retaining it for the nearly extinct Sea of Japan population. The IWC did qualify the gray's new status, restricting the quota of 178 to aboriginal hunters or to a government hunting in behalf of its aborigines. In fact, the IWC was simply recognizing the status quo, but the move was disturbing. The gray is the first protected whale species to be unprotected in the name of scientific conservation.

The gray whale has been through it all. Its North Atlantic population may have been the first stock whaled out of existence. It fell victim to the

Yankee pillage of the Pacific and to the renewal of whaling by catcher boats and factory ships. It qualified as the second species protected from commercial whaling and has since been whaled commercially because of the loopholes of aboriginal exemption, scientific permits, and the New Management Procedure. Now the gray whale is passing into another period. California has adopted the gray whale as the state marine mammal and has underwritten a project at Bodega Bay to record whale noises and to make some back in hopes of a reply. Now people go out to the headlands and down to Baja California to see the grays, to watch them. It used to be that we wanted to kill whales. Now we want to be their friends.)

Fear long stood in the way. Erle Stanley Gardner, eager to get away from his typewriter for a taste of adventure, trekked to Scammon's Lagoon in the late 1950s to check on reports that the whales were returning in number. He found them, and he found that their reputation among the Yankees as fierce fighters was well deserved. The whales, Gardner said, rushed his small boat again and again with undoubted hostility.

Ten years later, the first whale-watchers entering Scammon's Lagoon found the whales standoffish, but unaggressive. But the gray's reputation and the fear it engendered remained on everyone's mind. There had been mishaps. Paul Dudley White, the famous cardiologist, went to Scammon's to record the heartbeat of a whale with an electrode-harpoon. He never made the recording. After he startled a whale awake, no other allowed him close, and one rammed his boat and splintered the hull, almost sending White to the bottom. The Cousteaus, while making their film on grays, pushed one too far. Determined to get underwater footage of a swimming whale, two divers in a Zodiac chased the same animal for hours. Suddenly the whale came about, breached, and landed on top of its pursuers. The crash destroyed the rubber skiff, but the divers, by sheer good luck, escaped with only a few bruises and a sprained knee between them.

The watchword among the boat operators on whale-watching trips was, Never separate a cow from its calf. The cows protect their young quite fiercely, and getting between mother and young, with whales as with bears, was thought sure to draw a charge. To avoid such untoward encounters, the helmsman always stood in the stern and kept an eye out for cows and calves.

One afternoon during my first visit to San Ignacio, in February 1976, I was out in a skiff with five other passengers. The helmsman stood in the stern as we drifted motorless on the tide. He saw it coming first. He sat down fast, tried to start the engine, faced forward and braced himself. The whale surfaced not ten feet away, coming right at us. The head looked as wide as the skiff, and the animal's length was lost in its width. I tensed, seized the handline, thought to myself what I would do once I was in the water. Yet even as I was tensing, the whale's snout came deeply down, and the rolling back almost scraped the bow. The skiff rose on the wave the whale made as it passed underneath. The skiff dropped down, and I was holding my breath. I let it out in a chuckle, and everybody else was making that same ridiculous little giggle that follows on close calls.

Later the point sank in. We had avoided the accident only because the whale so wished it. It could have run right into the skiff and brushed it aside, like an elephant shaking off a fly. Instead, the whale took care to avoid collision.

That was but one of the auspicious occurrences in San Ignacio at the time. The week before, the *Finalista 100* had been visited at night by a juvenile whale that displayed great interest in the skiffs tied to the stern rail. It came up under them and rubbed its face against their sides and bottoms and tossed them about, something like a dog in a fantasy game with a bone. The crew called the whale George. George reappeared the week I was there, putting on the same display of batting the skiffs around. The crowd of passengers gathered, popping flashbulbs into the night. At certain angles, the light turned George's eye red, setting a momentary ruby in Leviathan's face.

Surprisingly, George put in a second appearance for us just after dawn, the first time he had come in daylight. Again the skiffs were the object of George's desire. He bumped and rubbed against them, scraping his jaws on the gunwhales. Again passengers crowded against the rail to watch and photograph. Every time George blew, there were cries of surprise and a mass turning away to shield camera lenses.

Bob, the deckhand, first got the idea. He pulled a skiff in by its painter and jumped into it. George came up to Bob's skiff and pushed it about. Mike, the captain, could not be outdone. He, too, went down into the skiff and rode the whale's tossings. Mike and Bob crouched down and hung on, but soon they grew used to the whale's movements and lost their fear. The huge animal rolled and lay still, and the two stroked his jaws. "I'll be the first to kiss a whale," Mike shouted. He chose his moment, leaned out, and smacked George full on his huge lips.

Apparently, George didn't mind. He sculled and lay flat on the surface so the men could stroke him. Now and then he left the skiff and circled the hull, but always he returned to Bob and Mike, as if asking for more attention.

George left for a time, but when he came back, he brought a friend. The new whale was cautious, following George in his circuits of the hull but

George lifting a skiff tied to stern of observer's craft

never coming close enough to be touched. As soon as it was obvious that this whale was becoming as much of a fixture as George, he required a name. It was Ralph.

From the upper deck, which gave a view far across the water, I could see that something was going on in the whole of the lagoon within eyeshot. The day before, the whales had stayed well off from the boat, but today some forty or fifty animals were congregating close by. To the port side, three whales moved about in a close-order group. They thrashed the surface with their flukes, then sprinted back and forth between shore and boat. It was a mating trio, one ready cow and two interested bulls. The whales rolled together side by side, and the bulls, their interest mounting to passion, traded leaps, throwing themselves from the water and falling back massively. Then a penis — pink as a new eraser, five feet long, curved and tapered — arced over the cow and probed her body. The other bull, in what seemed a remarkable lack of self-interest, held his body against the copulating pair to keep them from rolling apart until the deed was done.

George and Ralph stayed around the *Finalista 100* even when the skiffs were brought aboard in preparation for the return to San Diego. They followed the boat as it got under way, but as we picked up speed, they fell farther and farther behind. Finally they could be seen no longer. The people who had been watching turned about toward the open sea ahead.

The season still had weeks to run, more people came to San Ignacio, and George and Ralph changed from rarities to regulars. They visited the one or two excursions entering the lagoon each week of the remaining season, and George kept up his love affair with the skiffs, either empty or peopled. One of the tour boats carried George Bryant of the Toronto *Star*. His photographs of George's distinctly unlovely but undoubtedly large face went out over the AP wire and appeared in many newspapers. Unfortunately, the captions on the pictures, besides containing the usual newspaper errors and misstatements, had George's name wrong. They called him Nacho. Somebody — a source, usually reliable, describes the culprit as "real moldy; kept a bottle under the pillow" — changed George's name to Nacho, for the Mexican flavor no doubt. The newspapers liked that, and George became Nacho.

Whatever his name, George was the known first of a kind. Ted Walker, the trip naturalist and an expert on the gray, had touched whales before George, but they were always stranded babies, blistered by the sun, pecked by gulls, half-starved, and panicky with the approach of a death that proved inevitable. George was a creature of altogether different sort. He was a healthy juvenile male, probably four years old, and for whatever reason he was fond of people, that reason was his own. George was under no duress, in no trouble. He came because he wanted to.

In time, George lost his uniqueness. Other whales moved in on his act. The whales grew interested in people as their interest in sex waned. During January, when the number of whales in the lagoons is greatest, the bulls are hot and ever on the prowl. But some of them begin migrating slowly north about the beginning of February and by the end of the month the lagoon has calmed considerably. In the winter of 1977, as the mating died down, the whales evidenced more and more curiosity about people. They approached the big boats, often rubbing against the hull or anchor chain. They likewise came up to occupied skiffs, sometimes seemingly for a closer look, sometimes to massage their skins against the soft rubber. And the whales allowed themselves to be touched by humans. Apparently some of them came to like it, because they kept coming back for more.

I returned to San Ignacio in late February 1978. Touching a whale was no longer unusual. It had become the salt to the broth. "I've just got to touch one," one woman gushed to me before we left San Diego. "I'll just die if I don't." The whales were more than cooperative. One of the skiffs even came across two whales attempting to mate — with only two, they could only attempt. The whales postponed their courtship awhile and played with the skiff instead of each other, stroking the tiny craft with their flippers and staying with it for over an hour. Always the whales behaved carefully. Sometimes they push the skiffs around, but they have yet to capsize one or to knock a passenger overboard. Their flipper touches are true caresses, soft and delicate movements. The sensitivity of the huge animals, averaging forty feet long and perhaps thirty tons, is astounding. A whale lifted its head straight out of the water next to a skiff. A man aboard reached out and touched the underside of the whale's lower jaw with no more force than you would tap a friend's shoulder to get his attention. At the touch, a visible quiver ran through the whale's skin, and it dropped under the surface instantly.

As for the old saw about separating cows and calves, that has gone by the wayside. The helmsmen don't worry about it; the cows tend to their own as they wish. I was out one morning when the air was a dead calm. A storm was making its way in from the south, and under the still and overcast sky the water lay as thick and silken as quicksilver. A cow with a fairly large calf, probably about six weeks old, approached the skiff slowly. They circled several times, perhaps ten yards away. The cow was big, mottled with scars, and barnacled on head and neck. The calf was smaller, sleeker, too new at life to carry more than a few scars and freeloaders. The two stayed close, as if curious but uncertain of closing the last of the gap. Then the calf broke away and coasted slowly to us. It rolled onto its side and swam by, its belly perhaps six inches from the rubber skin of the boat. The cow surfaced five yards off the stern; when she blew, it rained on us. In my imagination I heard her calling the little one. He answered. She sounded, rose again in the same place, and the calf was with her. They blew together twice, mother and child, then made their way slowly down the lagoon toward a favored feeding area off Rocky Point.

Obviously, we are dealing here with a recurring phenomenon. George may have pioneered the movement, but it has clearly grown beyond him and taken in many whales. It demands an explanation.

One is that the whales are trying to "communicate" with us, that they are consciously and purposely breaking down the wall long standing between man and animals. You could call this the New Age interpretation. It implies that the grays are an alien, nonhuman intelligence that has consciously chosen, at this one point in all the millennia covering the history of whales and humans, to make contact with people for the first time.

The New Age version could be right, but there is no reason other than the notion's current popularity to believe that it is so. It is, New Age or Old, a long way from face-rubbing on gunwhales to the communication of wisdom. The explanation, frankly, is more ornate than the phenomenon. Other explanations bear looking at.

To begin with, gray whales like to touch and rub. They scratch themselves against fixed objects like anchor chains; mothers and calves touch constantly; lovemaking among grays entails a great deal of stroking and rubbing, most of it delicate and sensual. When a gray whale rubs against a skiff, it is not displaying some new activity or behavior. It is doing something it has long done; only the object, skiff instead of an anchor chain or another whale, is new.

The rubbing could serve a prosaically mechanical function. Gray whales are troubled by skin parasites, small crustaceans called whale lice (mostly *Cyamus scammoni,* named by the naturalist W. H. Dall as a scientific tribute to Scammon). Rubbing against an object dislodges the lice, and a springy skiff is a much more pleasant object to rub against than a chain or a boat hull. A few people in the skiff probably make it more desirable as a rubbing post. The extra weight increases the resistance and lets the whale rub all the harder.

This explanation probably accounts for at least some of the rubbing whales. But it doesn't explain the ones that caress skiffs for hours or the calves that coast by for a look. I asked Ted Walker what he thought. "Curiosity," he said. "Plain and simple curiosity."

Once the bulls leave and the food supply in the lagoon dwindles from all the feeding animals, there really isn't much for the whales to do except doze at midday and keep an eye, or an ear, out for sharks that may have designs on whale calf for dinner. The skiffs offer a welcome diversion. Gray whales are playful by nature; the calves frolic constantly, and the juveniles commonly enter the breaker zone during migration, apparently for the fun of surfing. They are also, like dogs that learn the joys of belly-rubbing, devoted sensualists. They come to people to be rubbed and touched, much as they come to the skiffs to rub themselves.

But why have the whales engaged in their come-to-the-people routine only of late? "Fear," Ted Walker said. "Our fear, not the whales'. People were too afraid of them to let them in close."

This could well be the heart of the matter. The New Age interpretation assumes that the whales have chosen this time to open channels of communication. We humans have been the same all along; it is the whales that have changed. But that simply isn't the case. The change has been ours.

When Erle Stanley Gardner went to Scammon's to see grays, he had good reason to fear the whales. The species had a reputation for aggression little better than the grizzly bear's. When a whale approached his boat, he assumed that its intent was hostile. But as years passed and more and more people saw the whales at close hand, it became apparent that the gray whale was not universally aggressive and dangerous. It had to be provoked. Finally people came to accept the animal, despite their innate and thoroughly reasonable apprehension about an animal so large, as generally friendly and well-meaning.

I do not for a moment believe that George came to the skiffs that night to teach us anything. But we're learning from him anyway.

A central part of meeting the whales in Laguna San Ignacio is meeting Ted Walker. He looks like Father Christmas, with a round face and a white beard, but he goes up the peaks and down the arroyos of the islands with the energy of a man twenty or thirty years his junior. That is only the first of Walker's many curiosities. He is walking testimony to a universe streaked with more anarchy and contradiction than science dare admit.

Walker came from Montana, found his way into graduate school in biology at Wisconsin during the days when Aldo Leopold was working out the practical ramifications of what we now call ecology. Walker was concentrating on cytology, the study of cells, when the Second World War broke out. He became a naval gunnery officer stationed on destroyers, a misassignment even by the navy's twisted standards. Walker gets seasick at the mention of the word *swell,* and his job took him up to the crow's nest, where the roll is the worst. But the navy did introduce the kid from the hinterland to the sea and to whales, which he saw for the first time during the Aleutian campaign.

When Walker returned to Wisconsin and his interrupted graduate work, he dropped cytology for limnology, the study of fresh water, and wrote his Ph.D. dissertation on the orientation of fish to aquatic plants. But Walker was already dissatisfied with limnology. Lakes and streams depend too much on the surrounding land; they are more a piece of a whole than a whole themselves. Walker wanted something big, something closer to truly self-contained. When Scripps Institution of Oceanography offered him a job working on the problem of the vanishing Pacific sardine, he took it. The job meant the sea, and the sea is big.

The sardine research was actually a reanalysis of somebody else's data, and to get his hand into real research at sea Walker joined the ongoing work on lateral lines in fish, then a chic topic in zoology. The experience alerted Walker for the first time to the constraints of orthodox science. When the navy offered Scripps a classified contract to come up with a way for distinguishing whale sounds from submarine noises on underwater listening devices, Walker joined the research team. From that time on, Walker worked almost exclusively on whales.

Scripps had become the center for scientific interest, limited as it was, in the gray whale. Carl Hubbs, now a professor emeritus at the institution, set up a whalewatch on the vantage point of Point Loma, and shortly after the war he organized the first aerial census of the Baja California lagoons. Walker and other Scripps scientists, Kenneth Norris among them, went to Mexico to observe the whales and learn something about them. Walker spent months and months with the grays. He developed a watcher's skills; he learned to look and to see precisely what was before him, not what he expected to see. Soon he was running against the conventional, and largely unsupported, wisdom about the gray whale. The story was that the gray whale did not feed in the lagoons. Walker saw the animals feeding. He set to work collecting evidence of feeding.

But all the while, Walker was becoming more and more disenchanted with the accepted scientific way of doing things. He came to see that many scientists, like many people, have mediocre minds, that they came up with mediocre ideas, and that the ideas became orthodoxy not so much because of their cogency as the political and institutional power of their authors. Science was dedicated to fine points, not totalities; becoming a scientist was entering a priesthood whose god was a tyrannical obsession with more and more facts about less and less. Ted Walker doubted that he belonged. He had little sympathy for what established science felt most sacred. And Scripps was having less and less sympathy for Walker.

Walker sought out new ways of making a living and working independently. The Cousteaus hired him as a consultant while they were making their film on the gray whale, and Walker spent two winters with them, mostly at the Boca de Soledad in Bahía Magdalena. Then he met a television producer who had an idea for a film series about the experiences of naturalists living alone in the wilderness. Walker jumped at the chance, he and Scripps had a mutually acceptable parting of the ways, and for six months he lived by himself in a cabin in southeastern Alaska. The idea for the series fell through, but Walker received the rights to the film he had made about his solitary venture. He went on tour narrating the film and followed it up with another about Baja California that ends on the high point of the lagoon whales. Now he also works as a naturalist for H&M

Landing, the San Diego company that runs the *Finalista 100* and the *Mascot VI*, another whale-watcher, and for Lindblad Travel.

The solitary experience in Alaska brought many things together for Ted Walker. His training and life had been a scientist's, a world of laboratories and libraries. Nature was always at arm's length, distilled behind the glass of a test tube or caught in the turgid language and sharp little numbers of scientific monographs. In Alaska nature looked Walker in the face. Nothing stood between them. Now Walker was like Thoreau, exiled by choice from conventional self-deceptions and reassurances and forced to deal with a baldly real world.

Science assumes itself to be dispassionate, unchanging; only the observed phenomenon changes. The thing seen is objectified. Walker calls this poppycock. When you watch something and objectify it, you no longer treat it as simply something other than yourself. It becomes instead something inferior. You strip the animal of what Walker calls its "inviolate right" and of any emotional or esthetic value. The animal's status changes from a being to be contemplated to a resource to be managed. Exit Thoreau, enter Rice and Wolman.

In this society scientists serve the same role as privileged interpreters that clerics did in the Middle Ages. An event occurred, the people went to the priest to find out what and why. Now we turn to the scientists, in all their variety from astronomers to zoologists. Scientific knowledge is rational-positivist knowledge, the sort spun from hypothesis and pegged to tests of evidence like experiment and multiple regression. Thus, to know an animal or a plant you must know its data. The flow of experience transforms itself into a line of cybernetic bits.

Walker explicitly rejects this view. He has no truck with "the naiveté of the assumption that the interpretation of an animal requires a scientist" — that is, that an animal can be appreciated only by knowing about it great heaps of minutial information. Such knowledge can actually get in the way of understanding and perceiving a creature as it is. You can know that a gray whale is sexually mature at six or seven, that the ventral grooves number two to four, that the mean weight for the species is given by the expression $W = bL^3$, but if you cannot sense the presence of the animal in the water — or, worse, do not permit that sense because it escapes measurement and mathematical expression — then you have lost much of the phenomenon you wish to observe.

"The whale is extraordinary," Walker says in the one evening lecture he gives about the whale on the *Finalista 100*, "not because of what we know about it, but because of what we don't know. Research on whales will challenge human ingenuity because underwater observations are necessary and we simply don't have them yet. So far it doesn't take much to be a whale expert. What I like best about these trips is that I'm never totally in command. I'm always finding out things about whales that I don't understand."

The Rice and Wolman study draws Walker's particular ire. Since science provides this society with its priesthood and its vocabulary of interpretation, we assume that whatever science approves must be correct because science approves it. If the orthodoxy of science approves a method and attitude of study and if that method is followed and that attitude maintained, then we accept the findings as true without thinking much more about them. The Rice and Wolman study adheres to the standard operating procedures of mammalogy. To study animals zoologists must have specimens. They kill some number of animals, then examine them for data to indicate facts about the living species from which the cadaverous sample was taken. We assume that since the evidence the cadavers provide is demonstrably material — it can be weighed and measured — then it must be evidence of the best sort.

Trouble is, there is no reason, apart from the standards of science, to assume that such evidence is better than any other sort — observations, for example — nor does the evidence, even if it is worthy, guarantee that the interpretation given it is any good.

To wit, cetologists have long held that migrating and wintering whales rarely if ever feed. The whale's annual life cycle is often described as feast changing to famine as summer becomes winter. There can be little doubt that migrant whales do get less to eat in the winter than in the summer. In the Antarctic, as one example of a feeding ground, the whales are thinner and yield less oil per ton of body weight at the beginning of summer than later in the season, after the long feed. But in much of the writing about whales the word *fast* is used as if some religious stricture bound the whales' great mouths tight and they, like the faithful on the Day of Atonement awaiting sunset and the first morsel, longed for summer and the groaning board of planktonic bloom.

There is evidence to support this picture of the gray whale. Scammon reported that he found in the stomachs of several grays taken in the lagoons only small amounts of a marine plant the whalemen called sedge or sea moss. Rice and Wolman examined the stomachs and intestines of all their 316 whales and found food remains in only two. They concluded: "While migrating, gray whales rarely attempt to feed, at least along the southern sector of their migration route. What little evidence is available also indicates that gray whales seldom, if ever, feed while on the winter grounds."

At the same time that Rice and Wolman and their assistants were standing hip deep in gurry and looking at carcasses at the whaling station in Point Richmond, Ted Walker was in Mexico watching the living whales. Observation may sound like the simpler job. Cutting up a dead whale is a ghoulish task, certainly less enjoyable than watching whales from an observation post under the good Mexican sun. But observation is absolutely the more difficult of the two. You have nothing to measure or weigh. You can only watch carefully in hopes of seeing what is truly there. And when you think you know how to explain what you are seeing, you have to support that conclusion as well as can be with information that is nearly as ephemeral as the act of watching itself. Watching is the art of science.

Ted Walker was watching and he noted several phenomena. To begin with, there was this business of what the whales were doing when they brought their heads straight up out of the water, held them above the surface for several seconds, then settled slowly underneath. The posture was often called spyhopping, from the assumption that the whale was putting its head and eyes up for a look around above the surface. Suspecting anthropomorphism, Walker photographed spyhopping whales, examined others' pictures as well, and found that in all but a small percentage the spyhopping whale's eyes were closed. If spyhopping whales were really spying, they didn't see very much.

Walker also noticed that the barnacle patches on the whales' heads were asymmetrical. One side was more thickly encrusted than the other. It looked almost as if the whales were rubbing off the barnacles on one side of the face and leaving the other undisturbed. Aerial films taken by Scripps scientists at Scammon's showed whales swimming through the shallows, diving, and raising a cloud of bottom sediment, repeating the action and leaving behind them a trail of great muddy florettes in a line slowly distorted by current.

Bit by bit, Walker built a picture of how a gray whale feeds on the lagoon bottom. The animal dives deep, bringing up its flukes as it heads down. On the bottom the whale rolls sidewise, almost like a partial revolution of a corkscrew, and scoops up a mouthful of mud. It may swim on, or it may lay its flukes into the trough it has excavated, put its back toward the current, and allow itself to be lifted up until the head clears the surface. This is the spyhop, or as Ted Walker prefers, the spar — not a way of looking around but of getting a meal down.

Bottom-feeding is only one of the ways the gray has of getting food, inside the lagoons or out. The whale also feeds on weak-swimming schoolfish, usually by dashing pellmell into the school with its mouth open and breaching from the water with the velocity of the rush. Often a whale feeding on fish will breach repeatedly at nearly regular intervals. Half a dozen leaps are common, and there are reports of forty. Pictures of such breaches show the feeding — water streams copiously from the mouth, which is sometimes open partway.

It is not really surprising that gray whales are seen feeding on the winter grounds, because these are rich areas. Bahía Vizcaíno, from which Scammon's Lagoon opens, is one of the most productive ocean zones in the world. The rest of the Baja California coast, while not that rich, is still a good feeding area, particularly because of the dense shoals of red crabs (*Pleuroncodes planipes*) found in these waters. It seems curious that wintering grays would be most abundant in the areas where feeding is the best if they did not feed.

Why then did Rice and Wolman find empty stomachs in their whales? "For one thing," Walker said, "if you looked at all the corpses coming into the New York City morgue, you'd find that a lot of them have empty stomachs. An empty gut doesn't mean much of itself. It certainly doesn't mean the whales fast all winter. All it means is that San Francisco is a lousy place to be a whale." Rice and Wolman, Walker thinks, mistook the atypical case for the general rule.

To us lay people, the debate over whether gray whales feed in the winter has little consequence. It points, though, to matters of considerably greater significance. Feeding is a straightforward, unequivocal behavior. A whale either feeds or it does not. But if the scientists are uncertain or mistaken about such basic matters, how far can their ideas on more crucial issues be trusted?

Consider one such crucial issue: population. Whale population data are important because they provide the supposedly scientific basis for whaling quotas. The gray whale serves as a good exemplar species on the question of population because the species's range is restricted and thus readily accessible to censusing from land, sea, or air. Or at least so it should seem.

It appears, from a painstaking historical analysis by David Henderson, that the population of gray whales before whaling began numbered 15,000 to 20,000. Whaling took a heavy toll. Given Roy Chapman Andrews's statement that the gray whale was extinct or nearly so by 1910, most writers assumed that the population comprised perhaps 1,000 or 2,000, a remnant cut further by the factory whaling of the 1920s and 1930s. In 1946 Carl and Laura Hubbs ran the first aerial census of Mexican wintering grounds and found only a few hundred whales. In 1952 the Hubbses made the count annual. By 1954 the total rose to 1,315 and in 1965 it stood at 1,581. Clearly the gray whale was recovering, albeit slowly. Rice and Wolman set the migratory population in 1969–1970 at 8,000 to

13,000, with 11,000 the most likely figure. There, they stated, the population was holding steady.

Rice and Wolman's estimate came from land counts made from two stations along the California coast and supplemented by offshore cruises to determine the percentage of whales migrating out of sight of land. These annual counts were continued after the Rice and Wolman study ended, and overflights of the migration route were begun. These data have recently been analyzed by Stephen Reilly, a doctoral student at the University of Washington and a colleague of Rice and Wolman at the National Marine Fisheries Service's Marine Mammal Laboratory in Seattle. Reilly's statistical study shows that the gray whale's population has been growing and now stands at almost 18,000.

All in all, the picture looks most hopeful. Whaling nearly destroyed the gray whale, but the few survivors bred the species back to prosperity.

But the numbers don't really work; there is a fly in the ointment somewhere. If the population numbered only a few hundred in 1946 and some 18,000 in 1980, then it grew voluminously. Stephen Reilly estimates the gray whale's net annual increase as 2.5 percent. If he is right and if the 1946 figure is also true, then the current population should be only 1,200, a good bit less than 18,000. Somewhere something doesn't fit.

Seiji Ohsumi, a cetologist with the Far Seas Fisheries Research Laboratory in Japan, put himself to the task of charting the gray whale's historical ups and downs. As Ohsumi sketched it out, the gray whale had an original population of something over 15,000. Aboriginal hunting through the centuries reduced the population to 11,600 by 1846, when the Yankees set upon the grays in the lagoons. With 8,000 whales killed by 1874, the population dropped to 4,400. This, Ohsumi maintains, was the gray whale's lowest ebb. Thereafter, the population increased slowly year by year, despite shore-whaling and factory ships. When Andrews proclaimed extinction for the gray whale, the California population actually numbered almost 8,000. About 1946, when the word was going around that only a few hundred grays were left, the true number lay between 9,000 and 10,000.

If Ohsumi is right, his picture is most curious. After Scammon and his fellows were done, 38 percent of the population alive when they began still remained, yet the whalemen themselves thought the species thoroughly depleted and not worth hunting other than haphazardly. Apparently, a whale species is commercially extinct well before it even nears biological extinction. In a way, this is good news, for it implies that some of the other commercially extinct species, particularly the finback, could be in better shape for the long haul than we realize.

But there is a deeply disturbing note here as well. It took the gray whale almost a century to return from its low point to the population level prior to Yankee whaling even though the surviving remnant was well over a third of that prewhaling number. Present estimates say that the worldwide populations of blue and humpback whales (12,000 and 6,750) are but 6 percent and 7 percent of their prewhaling originals. How long will it take them to recover? Indeed, with their populations so thoroughly savaged, can they ever make it back?

The situation is striking. Of all the great whales, the gray is absolutely the easiest and cheapest to study, yet our ignorance of these creatures remains monumental. It took Ted Walker's considerable energy just to overturn the orthodoxy that whales do not feed in the lagoons. And it has taken years of debate and research to figure out how many gray whales there once were and now are — and we still may not know for sure. Obviously we have precious little certain knowledge of whales, yet those who should be in the best position to realize this, the cetologists themselves, still call whaling "rational exploitation." The irony is as telling as the self-deception.

Greed poses one threat to the gray whale. It will take a revolution in the IWC to change the present rules, but such a change could come, and the whaling quota could be raised. This same scene has been acted out many times before. Always it leads to the same destructive climax.

There are other, more subtle threats to the gray as well. The exploitation of the petroleum of Alaska's North Slope could hurt the marine environment on which the gray depends in the summer. The same specter of oil pollution has raised its head in Mexico. Large deposits of oil and, particularly, natural gas underlie the Bahía Sebastián Vizcaíno and its peninsula. Pemex, the state petroleum monopoly of Mexico, has plans for various exploitation schemes, and at least one gas well has been sunk in Scammon's Lagoon, within the boundaries of the national park. A salt company called Exportadores de Sal has reshaped the tidal areas of the lagoon to make the world's largest drying ponds for the supersalty lagoon water. Barges move the salt through the treacherous lagoon entrance to a village on the south end of Isla Cedros, where it is offloaded for transfer to freighters. Such industrial operations in the gray whale's major breeding area, where the survival of the species from one generation to the next is vulnerable, pose a definite threat. For now, the whales are making do. But there is ever the prospect of catastrophe.

The bright spot in this jumble of scientific ignorance, corporate greed, and technological shortsightedness is the elevation of the gray whale to a creature of symbolic, even mythic importance on the West Coast. The gray whale, onetime dog food ingredient, is now a creature of high status. Hundreds of thousands of people lay claim to caring about the animal. The January parade of southing Leviathans draws them in droves to the headlands and cliffs. And every year a few hundred others take the time and spend the money to follow the whales south to the lagoons.

The question is whether all this human fluttering about the whales doesn't itself pose a threat to them. The land watchers, of course, make no difference to the whales. But boats at sea do.

On a calm January day in southern California, the sea seems paved with boats, mostly small private craft out for a look at the whales. The Marine Mammal Protection Act specifically forbids anyone in a boat from harassing a whale, but many boaters don't know or don't care about that provision of the law, which is hard to enforce anyway. Some boaters do harass whales, usually thinking that they can get a close look by running the whale down at high speed. This belief usually requires a big power plant, a dose of machoism untempered by reading *Moby Dick,* and your choice of either Jack Daniel's or Coors.

Gray whale experts like Carl Hubbs and Raymond Gilmore have worried that the frenzy could drive the whales farther and farther seaward. As it stands, there is no evidence that the whales are moving out. If they did, they would not be alone. About half the whales hug the southern California shore and expose themselves to the boating hordes, while the other half split off near Santa Barbara, pass the Channel Islands, and rejoin the coastal migrants near Ensenada and Isla Todos Santos.

At sea whales can readily move away, but in the lagoons they have no place to hide. Harassment poses a greater danger, one made more serious yet by the calves, which must be protected and defended by their mothers. This concern led the Mexican government to close Ojo de Liebre to all outside boat traffic. The only way to watch whales at Scammon's now is from the land or from the skiff of a local fisherman. This closing forced some of the San Diego boats out of the whale-watching business. The run to San Ignacio was too far and too expensive. As for those still making the trip — H&M Landing and Fisherman's Landing — the Mexican government has been picky about schedules and permits.

The feeling that whale-watching in the lagoons equals harassment is commonplace. Ted Walker once invited Joan McIntyre, head of the environmentalist organization Project Jonah, to come on one of the H&M trips aboard the *Finalista 100* as his assistant. McIntyre answered Walker's letter: "We have been advising people *not* to go see the whales while mating and courting because we feel that it is finally not in the best interests of the whales, and are working with friends in Mexico to try to reduce tourist impact on migrating whales."

In the winter of 1978, Steven Swartz and Mary Lou Jones, researchers from the San Diego Natural History Museum, set up a camp and an observation platform at Rocky Point inside Laguna San Ignacio to watch the whales and to watch the whale-watchers watching whales. Support for the study came from the U.S. Marine Mammal Commission. The point of the research was to see what the whales do inside the lagoons and to determine whether the boats make them change from whatever they are doing. When I was in San Ignacio in late February, the study was still going on, but Jones and Swartz both said that so far they had seen nothing to show that the skiffs or boats harass the whales.

My own experience in San Ignacio in Avons points up an obvious fact: the whales are in control. If a whale does not wish to be near a boat, it can easily avoid it. The only whales the skiffs get near are the whales that want the skiffs near. The helmsmen have found that it does little good to follow a whale in hopes of getting close. They let the whales take the initiative. Most of the whales are indifferent most of the time, but some of them, following George's long-standing example, seek the skiffs out to amuse themselves.

McIntyre's intent is good, but she has actually taken a stand against her own goal of preserving cetaceans. Watching the gray whales in Laguna San Ignacio works very much to the whales' benefit.

On the morning of the final day in the lagoon, I went ashore with a party of passengers to look for shells and tidal animals on the beach. The morning proved perfect. The sun burned off the fog and a fresh wind came from the land. In the bright clear light the lagoon channel looked emerald and the sand leading down to it was as yellow as a hound dog's coat. I settled down with my back against a dune and watched. The beach was good for combing, littered with remains and treasures, and the passengers were hunting intently. They went barefoot, pants legs rolled up. Beyond them, in the channel, the water was full of whales. They rolled and spouted, the plumes standing high and white in the sun. The whales rose up to spar and sank slowly down, and now and again one breached and came down massively. But the passengers walked on, eyes down, seeking sand dollars and moon shells.

"Now, that's jaded," I thought to myself. Two days before, when we had come into the lagoon, they had nearly climbed the deck rails in excitement, screaming and yelling and cheering at every single sighting of every single whale. Now, only forty-eight hours later, they were so used to the animals that they could comb a beach alongside them and not give the whales a thought. Human folly, I told myself.

It was later, on the voyage back home, that I realized that it had nothing to do with folly. It was a kind of wisdom, a widening of the world. When the passengers stood at the rail on the entry into the lagoon, they cheered because they were seeing something new and different. The whale was the great and unknown creature from way out at sea. They were now seeing it close up in what amounted to a spectacle. Over the next two days these people lived among the whales and actually touched them from the skiffs. They grew used to the animals, so used that by the end of their visit they could beachcomb with complete composure tens of yards from the same sight that had pumped them full of adrenalin only two days earlier. This was not ennui; this was acceptance. The whale had been unusual and alien. Now it was a proper, expected, and admired part of the world, as right as rain and as desirable.

The whalers have long had the protection of killing animals that few people knew and fewer cared about. Aldo Leopold called attention to this effect in *A Sand County Almanac:* "The erasure of a human subspecies is largely painless — to us — if we know little enough about it. A dead Chinaman is of little import to us whose awareness of things Chinese is bounded by an occasional dish of chow mein. We grieve only for what we know. The erasure of Silphium from western Dane County is no cause for grief if one knows it only as a name in a botany book."

And this is precisely what the gray whale has going for it. People do know it. Those who have been to the lagoons carry the experience within them. They are the gray's first and strongest line of defense. They realize that if the gray whale is lost, they too will surely diminish.

•   •   •   •   •

*When I examine myself and my meth-
ods of thought, I come to the conclusion
that the gift of fantasy has meant more
to me than my talent for absorbing positive knowledge.*
—A L B E R T   E I N S T E I N

The Jonah Factor

She was a handsome woman, dressed in caftan over slacks, pink on lavender. She faced a full auditorium, but she spoke in the soft, intimate tones in which one tells a few friends gathered close of meeting Jesus. Her contact had been personal, its effect certainly religious. "It was a little like falling in love," she said. On sensing the presence of the other, she felt "a crack in the cosmic egg, a tunnel of experience that scratched a genetic memory that I had experienced in some form." There was transcendence; there was even conversion. She came away "a fuller human being, not so arrogant."

The woman was Antoinetta Lilly, wife of John Lilly, and she was speaking not of a religious encounter but of swimming with a beluga whale. Until a few years ago, anyone confessing publicly to such an experience in such terms would have been considered a prime subject for psychotherapy based on the most unmentionable Freudian assumptions. But times have changed. On that November evening in Olney Hall at the College of Marin, people paid to hear Antoinetta Lilly, as well as her more famous spouse, and when she told of the ecstasy she felt as the beluga squirted her, not one person so much as giggled. They all took it quite seriously.

More and more people have been taking it more and more seriously. Conferences, symposiums, and celebrations of cetaceans keep popping up, and not just on the coasts. One of the earliest and biggest happened in December 1975 at the University of Indiana. Organized by John Jay Goodman, who knew nothing about whales but recognized a hot topic when he saw one, the National Whale Symposium filled the various halls, rooms, and galleries of the Bloomington campus with lectures, films, and art exhibits that brought together whale devotees of all scientific and activist stripes. Books, T-shirts, medallions, and posters of, on, or about cetaceans sell in all sorts of gift and nature stores. The demand is there and profits can be made, but woe to the moneyman who mistakes the public mood. No lesser magnate than Dino DeLaurentiis did. Wishing to cash in on the enormous success of *Jaws,* DeLaurentiis commissioned a screenplay by Luciano Vincenzoni and Sergio Donati and a spinoff novel by Arthur Herzog. Both versions of the story baldly copied *Jaws,* but replaced the shark with a killer whale. Despite the casting of Richard Harris and Charlotte Rampling in the film and the usual advertising hoopla, *Orca* proved a flop. People who had been happy to stand in line to spend money to be scared out of their wits by a shark refused to accept the same treatment from a killer whale.

And right there is a measure of how much the attitudes toward cetaceans have changed, of why no one laughed out loud at Antoinetta Lilly. In April 1954 *Time* magazine reported a good deed done by American GIs in Iceland. Local fishermen were complaining about orcas damaging their nets, and the military came to the fishermen's aid by machine-gunning the supposed pests. *Time* described the whales as "savage sea cannibals. . . . As one was wounded, the others would set upon it and tear it to pieces with their jagged teeth." Readers accepted such tripe in 1954, but two decades later the same public withheld their dollars from that same bloodthirsty image of the orca.

The truth of the matter is that whales are more than a little trendy. Time and again as I was writing this book, acquaintances old and new would ask me what I was working on. "A book on whales and dolphins," I would say, and the same response came back as if on tape, "Far out. I'm really into whales."

Into whales: a good metaphor, physically impossible and exactly symbolic of the new cetacean chic. The literal image puts a name to the present popularity of whales and dolphins: the Jonah Factor.*

---

* Restrictions on my poetic license demand the admission that the term is not my own. I read it in a magazine article within the past couple of years, but a long search failed to find the source. My thanks to that unknown author.

The Jonah Factor manifests itself in a widespread desire for information about cetaceans. Evidence of that desire abounds. Synergy Seminars of Sausalito, California, put together a "Celebration of Whales and Dolphins," an evening of films and lectures that played various California cities to full houses with only minimal advertising. At Fort Mason in San Francisco, sculptor and designer Larry Foster made a steel-framed fiberglass replica of a juvenile finback female he named Pheena. A helicopter carried the finished model to its place of display on the Marina Green, and the flying whale was filmed from another helicopter by a crew from *National Geographic*. When Pheena touched down, a hundred excited children descended upon her and jumped, scrambled, and shinnied up to sit on her black back. Byrd Baker, poet and commander of the Mendocino Whale War, put together a "crusade to save God's whales" in the form of a museum, packed it into a forty-foot bus, and set out on a meandering tour of the country. Project Jonah also has a whale bus, a VW van equipped in multimedia style to bring cetaceans to schoolchildren and community groups.

Some people have been trying for firsthand contact with cetaceans over longer periods and in other ways than the Baja excursions. Project Jonah is befriending a school of dolphins at an undisclosed location in Hawaii. Jim Nollman of Bolinas, California, puts on a wet suit and enters the fiercely cold winter waters off Point Reyes to serenade the passing gray whales with his mellifluous whale singer, an instrument he himself invented from Native American tone drums. Paul Winter, who plays excellent saxophone, and seventy other musicians and artists went to an island in Bahía Magdalena and made their music for the wintering gray whales.

There is a certain urgency to this desire for information and contact, an impatience borne of the knowledge that the whalers still take nearly 20,-000 whales a year, that tuna fishermen kill dolphins, and that for certain cetacean species extinction is a very real threat. People who learn about whales and dolphins are often drawn to do something to stop their killing, a commitment that is very much a part of the Jonah Factor.

The commitment to action against whaling reaches its epitome in Greenpeace. The organization is Canadian, created originally by people in British Columbia to oppose nuclear testing in the Aleutians and the South Pacific. Greenpeace opposes nonviolently, by putting itself between the evildoer and his object, making itself both a hurdle to be leapt and a witness to the evil. To stop the nuclear tests, Greenpeace sailed into the test zones. Stopping whaling meant separating whales from catcher boats.

The first Greenpeace voyage to save the whales began in April 1975, and its purpose was to find the Russian whaling fleet known to hunt sperm whales off the coast of northern California. The Greenpeace vessel was the *Phyllis Cormack,* a converted fishing craft. It took two long and tedious months at sea to locate the Russians, and when Greenpeace faced the whalers, the confrontation was short and to the point. The crew of the *Phyllis Cormack* noticed the carcass of an apparently undersize whale floating in the water and went to photograph and measure it. A Russian catcher boat bore down on the crew, the dead whale, and the *Phyllis Cormack* with its high-pressure fire hoses spraying salt water. Later, the Greenpeace members tried to stop a catcher boat from shooting at a pod of fleeing sperm whales by putting two men in a Zodiac between the catcher and the whales. The gunner fired over the skiff and killed a sperm cow just ahead.

That near-miss and the image of the Greenpeace members confronting the factory whalers on the high seas provoked a sharp public reaction. Only then did the Russians, as so often bullyboys to the core, realize what game they were really playing. The Russians thought the contest was a simple test of strength, will, and power on the high seas. In those terms they won, but the struggle was never in those terms. The Greenpeace vessel had come to bear witness, and in this age of high technology, witness translates into electronic media. Greenpeace understood this; everything that happened was photographed or filmed and distributed. Directly, the activists knew, they could save at best a few whales, but the media event created by their confrontation could loose waves of public opinion that might save thousands, perhaps even bringing whaling to a quicker finish as an industry. The Russians won with their muscle, but the outrage at their dangerous and disdainful tactics massed opposition against them.

Next year the word filtered down through the Soviet bureaucracy to the North Pacific whaling fleet. The Russians refused to play. Instead of coming up almost to the United States' 12-mile limit, as had been their practice, the Russians remained a full 700 miles offshore, sacrificing a good whaling ground in the hope of escaping Greenpeace. Greenpeace did find them after a long search that left their new vessel, the converted minesweeper *James Bay,* low on fuel and the crew low on water. The ship broke off contact to refuel in Honolulu, then located the Russians again. This time the meeting was very different. As soon as the Zodiacs neared the catcher boats, the catchers ceased pursuing whales. They would not kill for the benefit of Greenpeace cameras, knowing that every death flurry would turn that much more public opinion against them. Greenpeace tried a new tactic, dubbed Operation Asshole, which called for running a Zodiac up the stern slipway of the factory vessel *Dalnii Vostok* to block it. The Russians countered by pumping rivers of whale blood down the slipway, marking the factory ship with an appropriately grisly wake.

In 1977 Greenpeace added a second vessel, the *Ohana Kai,* and came at the Russians twice. The media were in ascendance this trip; the *Ohana Kai* carried an ABC film crew. The Russians cat-and-moused, and for a time a submarine shadowed the *James Bay.* Greenpeace harassed the whalers' killing, the whalers hunted as they could, and the now established stalemate remained largely in force.

Greenpeace's ethic is often adventurist and elitist, its tactics too often photogenic instead of political. Yet Greenpeace has done well by its mission. It has made a public issue of whaling's brutal stupidity, and it has forcefully made the point that the whales belong as much to those who would let them be as to those who would destroy them.

It is easy to argue against whaling in economic and ecological terms. The industry is too small and its products too easily replaced to justify the use of so unique a resource as the great whales. But there is another, more deeply philosophical element to many arguments against whaling and to the Jonah Factor, an element that makes cetaceans the object of far more public concern than any other creature, with the sole possible exception of the harp seal. That element is the sense that cetaceans are somehow less like other animals than they are similar to us humans.

The ultimate source of this feeling lies no doubt well back in our minds and mythology, evidenced by the story of Dionysus and the pirates-turned-dolphins. But the recent and most proximate cause is the work of John Lilly. Lilly writes in the opening chapter of his *Man and Dolphin:* "We must strip ourselves, as far as possible of our preconceptions about the relative place of *Homo sapiens* in the scheme of nature. . . . We cannot continue to insist that man is at the top and that no further evolution is possible. . . . There may be a species as high on the evolutionary scale as man. . . ."

This philosophical conviction has important ethical ramifications. It is one thing to knock a fish on the head and fillet it for supper. Somehow it seems proper to take lower life in the service of higher. But between equals, such action is reprehensible in the extreme. So, with cetaceans, the question of similarity is very much to the moral point. If they are our equals, we do a great evil in shooting them with cannons or snagging them in nets to smother. This is not rational exploitation; it is murder. And keeping cetaceans in tanks for public amusement or scientific experiment is not captivity; it is imprisonment.

Such sentiment occasioned the Case of the Liberated Dolphins, played out by Louis Herman, a psychologist at the University of Hawaii; Kenny LeVasseur and Steven Sipman, students and tank cleaners; and Puka and Kea, bottlenosed dolphins and Herman's research subjects at the university's marine laboratory at Kewalo Basin. Early on the morning of May 29, 1977, LeVasseur and Sipman removed Puka and Kea and released them into the sea at Yokohama Beach. In the tank where Puka and Lea had lived floated instead two rubber dolphins bearing the graffiti "Let my people go" and "Slaves no more." LeVasseur and Sipman owned up to their act. On a blackboard in the laboratory they wrote, "Went surfin' — Kenny, Puka, Steve and Kea. Aloha."

Puka was never seen again. But reports of a dolphin playing with swimmers alerted Herman and his assistant to Kea's location a few hours after her release. When they got to her, she was hurting. Her flank bore a bad cut, probably from hitting a reef, and one eye was swollen nearly shut. Hampered by a lack of proper equipment, Herman's team could not secure Kea. The men stayed with her all day as she swam close by them, often whistling distress calls, but at night the cold and word from the locals that the beach was a prowling ground for sharks drove the humans out of the water. In the morning Kea was gone.

LeVasseur and Sipman said that they had released the dolphins because Herman treated the animals harshly. Specifically, they said, he had denied Puka a Frisbee that was her most beloved plaything. Herman said he did that because Sipman and LeVasseur gave the Frisbee to Puka to sabotage his research. He also said that the two were incompetent. In fact, he had fired them only shortly before they released the dolphins. LeVasseur and Sipman did not agree that they had been fired; they said Herman had laid them off.

Shortly after they were indicted for felony grand theft, Sipman and LeVasseur justified their actions to the press. "We have strong feelings about the dolphins," Sipman said. "They are highly intelligent creatures and we have a strong emotional attachment to them. They are brave, strong, and free, and they should remain free." "We did not steal them," LeVasseur added. "We gave them back." Herman disagreed. He maintained that the animals were properly cared for, fed well, and attended to carefully and that the dolphins enjoyed their learning and perception experiments.

The contrast between the antagonists could hardly have been more marked. LeVasseur and Sipman were tall and blond, overage undergraduates brought to Hawaii by the sun and water, not the pursuit of academic excellence. They surfed. They wore long hair. To hear Sipman talk about dolphins in the sea, you would have thought he was describing a commune in the Oregon backwoods. Herman had nothing to do with communes, backwoods, or hippies. He was a go-it-alone scientist — in fact, the only American academic with a research interest in the mental workings of dolphins — an East Coast city kid by upbringing and temperament, a success by all the standard standards. The soft mysticism of the flower children had wrapped itself about the hard edge of scientific rationalism.

The court ignored the confrontation of life-styles and philosophies. Judge Masato Doi, who presided over LeVasseur's trial (which was separated from Sipman's), cared about nothing but the wording of the criminal code of Hawaii. He disallowed testimony on any matter outside the bare facts of the case. The result was straight law. By that law the dolphins were chattels, the property of Herman and his research unit. LeVasseur had taken that property. He was thus guilty of felony theft, and his sentence was six months' imprisonment.

In the strictest legal sense, of course, Doi was correct. But the defenders of LeVasseur and Sipman maintained that the right lay outside the law in an understanding of dolphins as human equals. Thus, this argument ran, the act was no theft, but a liberation, a civil disobedience based on a good higher than the law, a good the law itself trammels.

The trouble with this argument is that Sipman and LeVasseur botched their supposed liberation. Since they released the animals only after Herman had given them thirty days' notice of dismissal, their motives are suspect. It could be that they found a helpful excuse for an act of childish vindictiveness, that they would have acted differently had their jobs not already been forfeit. And, ultimately, the release was a thoughtless and ill-considered act. Puka and Kea were Atlantic bottlenosed dolphins, which are physically and vocally distinct from the Pacific bottlenoses found around Oahu. The resident dolphins would probably not accept them, at least not for a period of weeks, and in those waters the group is the principal protection against sharks. Releasing Puka and Kea probably doomed them to a violent and premature death. It is unfortunate that men who laid claim to such highminded purpose did such a simpleminded job of thinking out the practicalities.

John Lilly found himself in a curious position in the Case of the Liberated Dolphins. Clearly, Lilly authored the present egalitarian element of the Jonah Factor, and his own release of experimental dolphins at his laboratory on St. Thomas ("I no longer wanted to run a concentration camp for my friends") prefigured Sipman and LeVasseur's action — except of course that Lilly was letting his own property go, an exception very much to the legal heart of the matter. The Sipman-LeVasseur camp claimed Lilly as an authority. Copies of his article "The Rights of Cetaceans under Human Laws" were distributed outside the courtroom in an apparent attempt to influence public opinion. Lilly publicly repudiated this use of his work, refused to have anything to do with the trial, and issued a public statement that read in part: "Many of us working with cetaceans are trying to develop criteria by which scientific comparisons of intelligence can be obtained. Unfortunately, public controversy resulting from the illegal and revolutionary behavior of these men can act to prohibit the kinds of research that are essential, if this knowledge is to be obtained. If the proponents of freeing captive dolphins have their way, by law or otherwise, proper scientific research will cease."

In a way, Lilly's position is ambiguous in all of this. He is himself a scientist and appears to support work of the sort that Herman was doing, yet Lilly's writings provided the philosophical basis, such as it was, for Sipman and LeVasseur's action.

Although Lilly writes again and again that cetaceans are an alien intelligence and, therefore, necessarily different from us, he does little with that difference and instead treats the two forms, dolphin and human, as more or less similar. His volume of collected works, *Lily on Dolphins,* bears the subtitle *Humans of the Sea.* He could hardly equate humans and dolphins more explicitly.

When Lilly does point up differences, he finds ones curiously sensible in our terms. We humans are a miserable, contentious, brutish, and warring lot. This we all know. The philosophers call it the human condition; the theologians, original sin. Whichever name you prefer, it is the price we pay for our brand of intelligence, for the ability to imagine both good and evil and choose between them. Somehow the dolphins have escaped our miseries and slipped the bonds that hold us fast in our predicament. Recall Lilly's description of delphinid life to his workshops at Big Sur: "The dolphins were close. . . . They had a freely moving and joyful life constantly together and . . . they had no compunctions about bowel movements, urination, or their sex life. I intimated that humans could well afford to try this way of living. . . . Today I still feel that by emulation of the dolphins' ways we could make a lot faster progress in loving one another and in enjoying life and abolishing tensions. . . ." To boot, the odontocetes have put their brains to better use. We sense that there may be higher planes. The sperm whale already lives there: "The sperm whale probably has 'religious' ambitions and successes quite beyond anything that we know. His 'transcendental religious' experiences must be quite beyond what we can experience by any known methods at the present time. Apparently, we can in rare times with our experience begin to approach his everyday, accomplished abilities in the cognitive, conative, and emotional spheres."

Lilly is hardly alone in his view. A Project Jonah newsletter, as one example, says: "The kind of vacation we will study is called life. The way we will study it is to not study it. It is a mystery to be experienced. Experiencing a mystery without study is called 'PLAY.' Dolphins are the play masters. By doing what they do we will learn more about being what they are." Dolphins, it seems, are masters of Zen as well as life. Recently, Joan McIntyre, the head of Project Jonah, changed the organization's emphasis from the lobbyist politics of stopping whaling to experiencing the mystery of cetacean life. There is good ethical reason to such experience. The cetaceans can be models; they should be, since they are better than we are. Margaret Howe, Lilly's assistant who lived in a tank with Peter the dolphin, wrote in her diary: ". . . I am implying the possibility [in dolphin life] of a lack of wheeling and dealing, of cheating and stealing and lying and other seemingly small but nevertheless devious ways of life stemming from human foible. From what I have observed and felt I do not feel that a dolphin newspaper, if one could exist, would contain articles on robbery, murder, dishonesty, delinquency, riots. . . . Let us be open to the possibility of learning and practicing what we learn from the examples set by the peaceful, gentle, and not to be overlooked in a time when ulcers and nail-biting are part of our everyday life, *happy* dolphin!!"

Steve Sipman held the same position. He told Arthur Lubow of *New Times* magazine, "My own observation of dolphins in the wild is that they make love and ride the ocean waves. They spend most of their time playing." This, says Karen Pryor, is the dolphin as floating hobbit.

The question, of course, comes down in part to one of fact. Are the cetaceans really the only species free from the dictates of nature? Lilly is a poor authority for the answer to this question; his work and his speculation spring solely from work with captive dolphins. The few studies in the wild indicate that the dolphins live pretty much as other mammals do. Their schools are hierarchical and the males in particular fight for position in a pecking order. Some of those fights get rough. According to one estimate, about one of every six male dolphins entering puberty dies before adulthood because of fighting. For dolphins, as for humans, life can be poor, nasty, brutish, and short.

But the issue here is not merely empirical. The dolphin-as-floating-hobbit has spread along with that curious assemblage of groups and idea, from est to Silva Method, called the human potential movement, whose purpose it is to bring us to the New Age. It springs from a dissatisfaction, spawned by affluence, with the bourgeois mentality, which is in part responsible for that very affluence. The typical middle-class American is up-tight, status-conscious, money-grubbing, and sexually restricted. This style of life leads to prostate trouble, heart disease, and a harsh view of the world. It causes cancer, too. The New Age will arise with the liberation of the mind from these strictures and with the creation of an individual who is loose, easy, flowing, sexually experimental, centered, and able to be-in-the-moment. The cetaceans, so the New Age spokesmen intimate, have already hit upon this natural wisdom. Dolphins are not merely floating hobbits; they are flowing utopians.

This view of the Old Bourgeois versus the New Age as the whole spectrum of conscious and intelligent experience involves an incredible presumption. It smacks of the evangelical who reads the Old Testament and assumes that its prophecies of doom apply, of all the countless centuries of human civilization, to his own insignificant lifetime. That view demeans whatever God may be at work in the universe. The floating utopian demeans the rightful place of the cetaceans as other species. Instead of seeing them as different, distinct, and unique, we require them to be what we wish ourselves to be. We are not looking at the cetaceans; we are holding a mirror up to ourselves. The narcissism of this view is much like the standard run of science fiction, in which you journey to the edge of the universe and find, as in one episode of *Star Trek*, Chinese Communists fighting round-eyed Yankees who say the Declaration of Independence as a prayer. That reconstruction of our own thinly disguised forms in an alien place, be it a distant planet or the mind of the bottlenosed dolphin, is a failure of creativity, of art, of the gift for fantasy that gave Einstein his great and universal ideas.

Perhaps worse, there is little new in the New Age view of the cetaceans. It is only the anthropocentrism of the centuries with a contemporary twist. From the time of our earliest myths of special creation, we have seen ourselves as set apart from the other animals. We are lord and master; the others, the underlings, are less than we are. In the great chain of being, we were the first and strongest link. Darwin shook this comfortably lordly world by showing that our species like all others is the product of time and natural selection. But the point never quite sank in. The old human-highmost habits of mind remained. Evolution, which is chaotic and purposeless, became a synonym for perfection, which it is not. All living things evolve; we are the pinnacle of evolution. Much of the present resistance to the anthropological findings of the Leakeys in Africa, which show that several closely related prehuman species once shared the same turf, springs from the anthropocentric prejudice that we must prove to be the crowning glory of a single line of descent, not the sole survivor of what used to be a big family. Thus, we have taken our need for primacy and underpinned it with Darwin instead of religion. If God did not put us in the driver's seat, evolution did.

The New Age view retains this ancient anthropocentrism. We remain at the top; cetaceans are our equals; therefore, the cetaceans must be like us. Our high status remains assured; we simply must share it. This easy solution escapes the real challenge, to understand other species as others, not as lesser beings but as different beings. The discovery of language ability in chimpanzees and gorillas and of the complex social nature of wolves shows just how much our science has missed, how much we still have to learn about these others. The cetaceans remain one of the most poorly understood animal groups. Unraveling their reality will widen science and philosophy. But casting them as nature's carriers of our current fantasy of salvation and the good life will keep the old anthropocentric blindfold tightly tied.

The 44 volumes of the Count de Buffon's *Natural History* (published 1749–1804) laid part of the philosophical foundation on which Darwin erected his theory of evolution. Buffon's work was popular, however, not because of its philosophical prowess but because the count had a way with words: "What has ever caught the imagination more than the dolphin? When man traverses the vast domains his genius has conquered, he finds the dolphin on all the seas' surface. . . . Man sees him everywhere — light in his movements, rapid in his swimming, astonishing in his jumps, delighting in charming away by his quick and foolish movements the boredom of prolonged calms, animating the ocean's immense solitude, disappearing like lightning, escaping into the air like a bird, reappearing, fleeing, showing himself again, playing with the wild waves and braving the tempests. The dolphin does not fear the elements, nor distance, nor the sea's tyrants." Terror gave us the dragon, the chimera, and all the other monsters. The dolphin came from the other side, from gratitude and from gladness at the warm sun and the smooth caressing water. The dolphin shows "that Greek spirit for whom Nature laughed and land and wooded mountains and flowering valleys were animated by voluptuous games, by varied pleasures, indulgent gods, and inspiring loves. The genius of Odin and Ossian could not conceive of him [the dolphin] amid the dark winters of the polar countries, for if Nature's dolphin belongs to all climes, the poet's dolphin belongs only to Greece."

The cetaceans touch this deep and happy chord in us. Seeing dolphins springing from wave to wave or following a great whale in its stately progress through the sea sears the mind, burns in an image of life exuberant and huge. That sense is very much to the heart of the Jonah Factor.

The jeremiads of ecological doom ring all around: poisoned food, impending extinctions, depleted ozone, carcinogenic drinking water, altered landscapes, and shrinking rain forests. The cetaceans have become barometers of the world's environmental health. Their survival and success means that there is a good measure of hope, that the oceans are in reasonable shape, that international marine policy and law employ some measure of wisdom. And their demise would mean the end of us for sure. If the cetaceans go, we can number our years.

But the cetaceans are more than indicators. They are symbols, bearers of meaning, as Buffon saw. They say something about life, particularly its glories.

About twenty miles from the house where I am writing these words there is a good place by the sea. You walk out to the end of a broad peninsular bench two or three hundred feet above the water till you reach the end, where it seems as if you are at the very edge of the earth. To both sides and ahead is the sea, a roiling confluence of the wave current from the northwest and the eddy from the bay. This is a good place to go in the spring, a good place to sit and watch. Everything required is present. There are land, air, sun, sea, and whales.

The whales are grays coming up from Mexico toward the Alaska summering grounds. They travel close alongshore, follow the curve of bay, and lie in the confluence to rest and feed before bucking the swells to round the point. This northward migration lacks the numerical spectacle of the southward. The whales journey over a much longer period and they are spread well out, a few today, perhaps a few more tomorrow, sometimes none at all. This is just as well. It makes the event less a circus than the closing of a circle. And it is just as well that you may have to sit and wait for the whales. As time passes, you cease to be a visitor and become a piece of this world. When the whales come, they are as much of a piece as you are.

Yet the first sight always takes you a bit by surprise. One moment there was empty water, all swells and riffle, and the next there is a broad and blotchy gray back and the shower of a spout. Another back and another yet and more spouts. The three whales, two adult cows and a four-month-old calf, come slowly along the base of the cliff and head out into the confluence, its crisscrossing swells as rough as the face of a grater. An hour they spend there, the cows diving deep and long to feed on the bottom, the calf nursing briefly from its mother as she lies on her side with a

Breaching gray whale and companion

flipper in the air like a spar. Then they turn north and west and the essence
of this event, which now seems normal to your eyes and mind, begins. The
swells come head on and pound against the cliffs, filling the water with
noise that keeps the whales from hearing their way through the passage.
To see where they are, the cows breach in turn, flinging their great bodies
from the water in a shower of spray and coming down in a huge splash.
Five times each cow breaches, ten leaps in all, the water shining silver in
the sun. Then the whales round the point and are into the wide sweep of
the Pacific beyond, out of eyeshot.

This is life served up as grandly as it ever is, huge and pulsing. It makes
you shed your hard and rational edges, conjure all the childishness, the
insanity, the foolishness, and the mysticism you have hidden away inside
you. The spirit of godly gamesomeness Melville called it, and you have it
bad.

And you realize, in the midst of that moment, the silent horror of a
world empty of whales.

A P P E N D I X

＊　　　　＊　　　　＊　　　　＊　　　　＊

The Cetaceans, Specifically

Aristotle, the founder of so many things intellectual, founded taxonomy. He recognized that animals could not be classified by a single characteristic. Putting all the flying animals together produced an obviously disordered agglomeration of grasshoppers, beetles, birds, butterflies, and bats. Instead, Aristotle classified organisms by many characteristics, by the plan he saw underlying their structure. He divided the animals into two basic groups, corresponding to vertebrate (with a backbone) and invertebrate (without a backbone), and then subdivided these groups.

Taxonomy, like most branches of natural history, declined after Aristotle, and it was not until the Renaissance uncovered the Greek legacy of thought and the exploration of the Americas sent a stream of unknown plants and animals to the study tables of European naturalists that scientific interest in classification was rekindled. One such interested party was John Ray, an English clergyman of the late seventeenth century, who catalogued over 18,000 plants and developed the first thoroughgoing classification of the animals. More important, Ray defined exactly what taxonomists were dealing with. Obviously, cats differ from dogs and the two constitute separate kinds, but are white cats and black cats likewise different kinds? Ray used the word *species* (Latin for "appearance"), defining it as a morphologically distinct type that can reproduce itself. Ray's definition, which made the crucial link between species and heredity, conveys the current meaning of the word.

The greatest taxonomist of all was Carolus Linnaeus (1707–1778), who succeeded at his seemingly impossible life goal of classifying all the known plants and animals. Linnaeus ordered the system of classification and made it useful for every-day scientific work. Naturalists recognized already that similar species could be lumped into a more inclusive group, which they called a genus (the plural is *genera*). The dog, jackal, coyote, and wolf belong to the genus *Canis;* the various species of catnip to *Nepeta.* Prior to Linnaeus, each species bore a phraselike name conveying its distinguishing characteristic. For instance, one of the catnips bore the appellation *Nepeta floribus interrupte spicatus pedunculatis* (catnip with flowers in a disconnectedly stalked spike), not the sort of name one dropped in a casual botanical aside to one's ladylove on a stroll through the garden. Linnaeus simpli-fied the system. To the generic name he added a single distinct shorthand label for the species, *cataria;* thus, *Nepeta cataria.* The Linnaean names caught on quickly, particularly among scholars tied yet to quill pens, and they remain in use today.

Linnaeus believed that the taxonomic system elaborated the acts of divine will behind God's creation of the world. The theory of evolution changed the meaning of taxonomy. Contemporary scientists see classification as a clue to evolutionary history. Closely allied species diverged from common ancestors fairly recently; more distant groups took separate paths farther back — the more distant the groups, the farther back the fork in the road. The ladder of taxonomic groups points to the historical progression of evolution.

Animalia	Kingdom
↓	↓
Chordata	Phylum
↓	↓
Vertebrata	Subphylum
↓	↓
Mammalia	Class
↓	↓
Eutheria	Subclass
↓	↓
Cetacea	Order
↓	↓
Mysticeti, Odontoceti	Suborder

The most fundamental division of organisms is into the five great kingdoms: Monera (blue-green algae and bacteria), Protista (protozoans, golden algae, dia-toms, and so forth), Fungi (fungi), Plantae (plants), and Animalia (animals). Whales and dolphins are, of course, animals. All in all, there are some 10 million animal species, covering a vast range of complexity and organization from the sponges to the mammals. Whales belong to the phylum Chordata, which means that they have a nerve cord on the back side and gill slits in the throat at some point during development. Altogether there are about 45,000 chordate species. Most of them, about 43,000 species, fall into the subphylum Vertebrata because their nerve cords are encased in a rigid column of bone and because they have a skull enclosing a brain. The vertebrates are then divided into seven classes: Agna-tha (lampreys and hagfish), Chondrichthyes (cartilaginous fish like sharks and rays), Osteichthyes (bony fish), Amphibia (frogs, toads, and salamanders), Reptilia (lizards, snakes, and turtles), Aves (birds), and Mammalia (mammals). Whales have all the mammal characteristics: warm blood, milk-secreting glands on the female, two-boned lower jaw, three-boned middle jaw, and seven bones in the neck region of the spine. The mammals fall into three subclasses: Prototheria, the egg-laying mammals, or monotremes, like the platypus; the Metatheria, the pouched mammals, or marsupials, like the kangaroo and opossum; and the Eutheria, or placentals. The placentals comprise twelve orders, of which one is Cetacea — whales and porpoises and dolphins, aquatic mammals without hindlimbs. The Cetacea are further divided into two suborders: the Mysticeti, or baleen whales, and the Odontoceti, or toothed whales.

The cetacean species now inhabiting the world's oceans, rivers, and lakes by no means represent all the forms that have existed. The whales and dolphins have been around in one form or another for the better part of sixty million years, and in that considerable length of time many species have arisen, flourished for a time, evolved perhaps into other forms, and finally vanished with no trace but a few cryptic fossils. Those few fossils have given paleontologists evidence for their picture of the likely course of cetacean evolution and provide part of the rationale for the classification of cetaceans now in use.

What follows is a layman's catalogue of the contemporary cetacean species. No taxonomy is ever absolute, and in this case, with knowledge of the cetaceans still largely scanty, various details of cetacean classification are the subject of continuing, and probably endless, scientific debate. This survey follows the groupings and scientific nomenclature of Dale Rice's *List of the Marine Mammals of the World* (NOAA Technical Report NMFS SSRF-711; April 1977).

SUBORDER MYSTICETI — BALEEN WHALES

Baleen whales have baleen filter plates and no teeth, symmetrical skulls, and two blowholes. The living species number ten.

FAMILY
Eschrichtidae — Gray Whales

Eschrichtid whales have no dorsal fin, two to four throat grooves, short and coarse baleen, and small heads compared to other mysticetes. The name Eschrichtidae is zoological homage paid to D. F. Eschricht, a nineteenth-century Danish cetologist.

Genus: *Eschrichtius*

Eschrichtius robustus — gray whale, musseldigger, ripsack, hardhead
The gray can be recognized in the field by its grayish color, often blotched with scars and barnacles; its lack of a dorsal fin; a ridge of knobs or bumps on top of the tail stock in front of the flukes; and a low bushy blow. At adulthood the gray runs from 40 to 50 feet long. The gray frequents shoreward water, regularly entering bays and lagoons. Historically there were three stocks: North Atlantic, western Pacific (Kamchatka-Japan), and eastern Pacific (Alaska-Mexico). The North Atlantic population has been extinct for at least 150 years, and the western Pacific pop-

ulation is either extinct or reduced to a few dozen. The California stock, probably numbering 17,000 animals, alone remains.

FAMILY
Balaenidae — Right Whales

The right whales have stout bodies, large heads, no throat grooves, and long, narrow, elastic, and very dense baleen plates.

Genus: *Balaena*

Balaena glacialis — black right whale, right whale, Biscayan right whale

The right whale has a long and slickly black body up to 60 feet long, a distinct V-shaped blow, and a head marked with white crusty patches called bonnets or callosities. The right inhabits temperate waters of the North Atlantic, the North Pacific, and the Southern Ocean. It migrates seasonally toward the polar edge of its range in the summer and back toward the equator in the winter. Heavily hunted in the past, the right whale is now rare and endangered. The estimated population is 3,250.

Balaena mysticetus — bowhead, Greenland right whale, Arctic right whale

The bowhead gets its name from its great arched mouth set in its equally great head, which may take up as much as one-third of its total length of 55 to 60 feet. The bowhead lacks the callosities of the right. An inhabitant of the floating ice of the Arctic, the bowhead migrates seasonally as the icepack advances and retreats. Four populations have been recognized: Spitsbergen–east Greenland; James Bay–Davis Strait–Baffin Bay; Sea of Okhotsk; and Bering-Chukchi-Beaufort seas. Only the last of these four stocks comprises more than a few hundred animals; its latest estimate is 2,300. The bowhead has the smallest population of all the great whales and is the most endangered cetacean species.

Genus: *Caperea*

Caperea marginata — pygmy right whale, broad-backed whale

Little is known about this curious animal. Growing to only twenty feet, it is considerably smaller than its two cousins, more streamlined, and equipped with the dorsal fin they lack. The pygmy right inhabits temperate waters of the Southern Ocean and is known less from field observation than from the remains of stranded individuals. A Soviet study indicates that this whale may be mistaken for the minke at sea, a confusion that could overstate the population of minkes and understate that of pygmy rights.

FAMILY
Balaenopteridae — Rorquals

The balaenopterids are the top-of-the-line mysticetes, the most recent evolutionary elaboration on a thirty-million-year-old design. Instead of the large mouth and long baleen of the rights, the rorquals have expanding throats plated with 40 to

100 grooves from which their common name comes — *rorqual* is the French form of a Norwegian word meaning "tubefish." Rorquals are relatively streamlined and fleet.

Genus: *Megaptera*

Megaptera novaeangliae — humpback whale

The humpback differs much from the other rorquals. It is thicker-bodied than the rest and a slower swimmer. The back is black, with a scraggy dorsal fin that in some individuals looks almost misshapen. Knobs and bumps mark the head, and barnacle colonies are often numerous and large. The flippers are long, white, and scalloped along their leading edge. Loafing humpbacks often beat the surface with their long winglike flippers. The humpback is playful and known for its plaintive songs. Found in both northern and southern hemispheres, this species typically winters and calves in the tropics and summers poleward. An easy target for the first mechanized whalers, the humpback now is rare and widely scattered. In the Atlantic, humpbacks are found in the winter off Bermuda, and they summer in the Davis Strait. In the Pacific, a few are found off central California, and a summering group appears every year in Glacier Bay, Alaska. The protected roadstead off Lahaina, Maui, is a well-known wintering ground for humpbacks that has become something of an attraction for tourists to Hawaii. Present global population: 6,750.

Genus: *Balaenoptera*

Balaenoptera musculus — blue whale, sulfurbottom whale, Sibbald's rorqual

The blue whale is the largest rorqual, the largest cetacean, and the largest animal in the history of the earth, regularly exceeding 80 feet and sometimes growing close to 100 feet and beyond. The blue whale's skin is distinctly blue-gray, sometimes tinged yellow with a film of clinging diatoms. The blue looks extremely large in the water, and its dorsal fin, set far back on the body, seems all too small for so gargantuan an animal.

The blue largely inhabits open ocean, and this habitat, combined with the animal's rarity, makes the species hard to find. The North Atlantic population is very small. Wintering North Pacific blues appear occasionally off California from Monterey Bay south to Baja California. Most of the remaining blues in the world are to be found in the Southern Ocean, particularly in their summer congregations along the Antarctic shelf ice. Although commonly thought to be nearly extinct, the blue whale appears to be surviving despite the fearsome damage done to the species before protection in 1966. The total population is now estimated at 12,000.

Balaenoptera physalus — finback, fin whale, herring hog

The finback is the second largest rorqual, growing to about 80 feet. It is gray-black above, white below, and is the only mammal colored asymmetrically; the right side of the head, lower jaw, and baleen are white, the left side black. The purpose of this unique pattern is uncertain. One hypothesis has it that the finback herds herring into a tight school by circling to the right about them and showing its light side, then turns into the packed fish and engulfs them. The dorsal fin of the finback is, as the name implies, prominent and distinct, formed into a sharp hook that shows above the surface as the animal bends forward to dive. The finback almost never shows its flukes when swimming or diving.

The finback occurs in all oceans, but rarely enters either the tropics or pack ice. The species adapts readily to different ocean environments, feeding on either small schooling fish or on krill. The finback was once the most numerous rorqual, but whaling has reduced its numbers to about 100,000 worldwide.

Balaenoptera borealis — sei whale

The sei got its name from a belief of Norwegian fishermen that this rorqual migrated to their shores in the company of the European pollack, a fish they called the sei. The sei whale occurs throughout the temperate waters of all oceans and is an even more eclectic feeder than the finback, foraging both by the usual rorqual method of engulfing and also by skimming the surface for fish, krill, squid, amphipods, and copepods. The sei typically reaches 50 feet or a bit longer. The body is blue-gray, the belly white or pinkish toward the tail. The baleen is dark with a white fringe. The sei has been exploited heavily of late, and the present population is thought to be a little less than 70,000.

Balaenoptera edeni — Bryde's whale

The Bryde's whale is so similar to the sei that it was long considered the same species. It frequents warmer water than the sei, regularly inhabiting the tropical Atlantic, Pacific, and Indian oceans. It stays closer to shore than the pelagic sei and is somewhat smaller, running about 45 feet long. The Bryde's can be distinguished from the sei at sea by a pair of ridges from the tip of the snout to the blowholes. The Bryde's whale feeds on fish and has much rougher baleen than the other rorquals. The exact size and distribution of the various stocks of this species remain to be determined. It is exploited off western Africa and in the Antarctic.

Balaenoptera acutorostrata — minke whale, lesser rorqual, little piked whale

The minke whale is the smallest rorqual, with even the largest females reaching only 33 feet. The minke's size is in part responsible for its name, which comes from the surname of a Norwegian gunner of crossed stars who killed a minke in the marvelously mistaken belief that he was blasting a blue. The minke has an evanescent or invisible blow that makes it difficult to spot at a distance, but it often approaches boats boldly and swims close by, where it can be identified by its size, its yellow-white baleen, and, in the Northern Hemisphere, by the white band on the flippers. Minkes in the Southern Hemisphere commonly lack the band. The minke occurs worldwide, but only rarely in the tropics, and is found commonly along the West Coast, being abundant inside Puget Sound. Until recently whalers did not bother with the species, but it is now hunted for meat, particularly in the Sea of Japan and off Brazil. The current world population probably exceeds 200,000.

SUBORDER: ODONTOCETI — TOOTHED WHALES

Toothed whales have one blowhole and, unsurprisingly, teeth. Most have asymmetric skulls. This accounting gives 66 species, with other lists running to 80 forms.

FAMILY
Platanistidae — River Dolphins

The river dolphins, because of their skeletal similarity to certain fossil dolphins, are considered primitive. However, the evidence is equivocal, and this family may be further divided. River dolphins have long thin beaks with 150 to 200 teeth.

Genus: *Lipotes*

Lipotes vexillifer — *pei ch'i,* whitefin dolphin, white flag dolphin, Chinese lake dolphin
   Known only from the Yangtze River and its tributaries, including Tung-T'ing Lake. Information about the animal is minimal.

Genus: *Inia*

Inia geoffrensis — bouto, Amazon river dolphin, Geoffroy's dolphin
   An inhabitant of the Amazon and Orinoco river basins of South America, the bouto reaches about 6 feet in length. Like the other river dolphins, it is social, generally living in small pods of a dozen or fewer.

Genus: *Pontoporia*

Pontoporia blainvillei — La Plata dolphin, *franciscana, tonina*
   A small dolphin found in the mouth of the Río de la Plata and adjacent coast of Argentina and Brazil. Certain scientists believe this species belongs closer to the ocean dolphins than the river dolphins, and it may end up as its own family.

Genus: *Platanista*

Platanista gangetica — Ganges susu, Ganges dolphin
   The susu grows to about 8 feet, is curious about humans, and despite small and rudimentary eyes is totally blind. It feeds by echolocation. The species occurs throughout the Ganges-Brahmaputra-Meghna rivers of south Asia.

Platanista minor — Indus susu, Indus dolphin
   Similar to the Ganges species, but an inhabitant of the Indus river system.

FAMILY
Delphinidae — Dolphins

The dolphins have a dorsal fin, 10 to 150 teeth in both jaws, notched tails, ungrooved throats, and often a distinct beak, which in some species is obscured by a large melon.

Genus: *Steno*

Steno bredanensis — rough-toothed dolphin

The rough-toothed dolphin looks something like the more commonly known bottlenosed dolphin, but it can be distinguished by its gradually sloping forehead and the yellow-white spots on the purple-black of its sides. The species is nowhere common in its range through tropical and warm temperate seas. Experience with the animal in captivity shows it to be quick-witted, attentive for long periods, and venturesome.

Genus: *Sousa*

Sousa chinensis — Indo-Pacific humpback dolphin, white dolphin

Found in the coastal zone of the Indian and Pacific oceans from South Africa to southern China, Indonesia, and Australia.

Sousa teuszii — Atlantic humpback dolphin

Found off west Africa, this species is distinguished from *S. chinensis* solely by tooth count.

Genus: *Sotalia*

Sotalia fluviatilis — *tucuxi, boto* (Brazil), *bufeo* (Peru), *pirayaguara*

Found in the Amazon River.

Sotalia guianensis — *tucuxi*

A coastal dolphin of northeastern South America. Some authorities believe that this is the same animal as *S. fluviatilis* and that the two species should be lumped into one.

Genus: *Tursiops*

Tursiops truncatus — bottlenosed dolphin

This is easily the best known of all dolphins, thanks to *Flipper* and the popularity of oceanariums. The animal can grow to 10 or 12 feet long and weigh perhaps 900 pounds, but 8 feet and 300 pounds or less are more common. The beak is as prominent and bottlelike as the name indicates, and the overall color is gray. Bottlenoses appear to swim slowly and deliberately, without the quick speed of some of the ocean dolphins. The species ranges widely in temperate waters of both hemispheres and occurs in greatest abundance close to shore and around shoals, islands, and banks. It is very much at home in shallow water. The species varies considerably in size and color pattern from region to region, and some taxonomists think that the bottlenoses from the Pacific and Indian oceans should be classified as the separate species *T. gilli* and *T. aduncus*. The bottlenose adapts well to captivity and learns quickly. In United States waters, bottlenoses are most common in the shoal waters of Florida.

Genus: *Stenella*

Stenella longirostris — spinner dolphin
Occurring in the tropical Atlantic, Pacific, and Indian oceans, the spinner gets its name from its characteristic twisting leap. It is slender, has a particularly long beak, and varies considerably in form and color. William Perrin (see Chapter Five) separates the spinners of the eastern and central Pacific into four subspecies. This animal typically forms large schools numbering in the thousands, and it often mixes with spotted and common dolphins.

Stenella attenuata — spotted dolphin, spotter dolphin
The spotted dolphin has a shorter beak than the spinner. Its body is gray with splotches of light gray and white that vary with maturity and from one geographical region to another. Perrin divides the spotters of the central and eastern Pacific into three subspecies. The spotted dolphin feeds principally on squid and runs in large schools. It occurs in the tropical and warm temperate waters of both Atlantic and Pacific.

Stenella plagiodon — spotted dolphin, Atlantic spotted dolphin
Found in the tropical and warm temperate waters of the Atlantic, where its range overlaps with that of *S. attenuata*.

Stenella coeruleoalba — striped dolphin, blue-white dolphin, streaked dolphin, long-beaked dolphin, Euphrosyne dolphin
The striped dolphin is large (up to ten feet in length) and beautifully marked with a pattern of black lines striping its dark back and white belly. The species occurs in temperate and tropical waters, but is rare along the North American coasts.

Genus: *Delphinus*

Delphinus delphis — common dolphin, saddleback dolphin
This abundant species, found throughout warm temperate and tropical waters, is strikingly marked in dark on light and elegantly shaped. It will ride the bow of boats devotedly. The common dolphin appears exuberant and playful at sea, but it does poorly in captivity, probably because its offshore habitat ill suits it for confinement. The common dolphin occurs off both American coasts. Some authorities believe that the Pacific group is actually two species, *D. delphis* along California and *D. bairdi* off Baja and in the Sea of Cortés.

Genus: *Lagenodelphis*

Lagenodelphis hosei — shortsnouted whitebelly dolphin, Sarawak dolphin, Fraser's dolphin

Inhabitant of the tropical and warm temperate Indian and Pacific oceans about South Africa, east Asia, and Central and South America.

Genus: *Lagenorhynchus*

Lagenorhynchus albirostris — whitebeak dolphin
Found in the North Atlantic from the Barents and North seas westward to the Davis Strait. This species, like the other members of the genus, is small (no more than 9 feet) and delicate in form.

Lagenorhynchus acutus — Atlantic white-sided dolphin
An animal of the North Atlantic from New England and Greenland to Britain, this dolphin is readily identified by its sloping forehead and prominent grayish sides.

Lagenorhynchus obliquidens — Pacific white-sided dolphin, hookfin dolphin
Found along the shores of both the eastern and western Pacific, the Pacific whiteside has the habit of riding boats on the stern wave rather than the bow. This species succeeds in captivity.

Lagenorhynchus obscurus — dusky dolphin
This species occurs in temperate coastal waters of New Zealand, southern Australia, Kergeulen Island, and South Africa.

Lagenorhynchus australis — blackchin dolphin
Restricted to southern South America and the Falkland Islands.

Lagenorhynchus cruciger — hourglass dolphin
Marked and colored so that its body suggests the intersecting parabolas of an hourglass, this handsome dolphin is found in the pelagic reaches of the Southern Ocean, most commonly just north of the Atlantic Convergence.

Genus: *Cephalorhynchus*

Cephalorhynchus commersoni — piebald dolphin, Jacobite, Commerson's dolphin
Coastal reaches of the southern tip of South America and the Antarctic islands and outliers.

Cephalorhynchus eutropia — black dolphin, Chilean dolphin
Restricted to the coast of Chile from about Chiloé Island almost to Cape Horn.

Cephalorhynchus heavisidei
Found along the southwestern coast of Africa.

Cephalorhynchus hectori — pied dolphin, whitefront dolphin
Occurs only in the coastal waters of New Zealand.

Top: Northern right whale dolphin (*Lissodelphis borealis*)
Bottom: Southern right whale dolphin (*Lissodelphia peronii*)

Genus: *Lissodelphis*

Lissodelphis borealis — northern right whale dolphin

This is a thin, handsome dolphin, dark above and white below. The right whale dolphin is no way related to the right whale, but like that species it lacks a dorsal fin. This species is probably pelagic, since it is sighted only occasionally in the coastal reaches of its Pacific range from Japan and the Kuril Islands to British Columbia and California.

Lissodelphis peroni — southern right whale dolphin

This species differs from the northern variety only in how far up the flanks the white belly color extends. The northern and southern forms may in fact be only geographical variations of a single species. The southern form occurs in the temperate waters of the Southern Ocean.

Genus: *Grampus*

Grampus griseus — whitehead grampus, grampus, gray grampus, Risso's dolphin

This large dolphin (up to 13 feet) is born dark, but whitens noticeably as it ages. The dorsal fin, which is prominent, and the flippers remain dark. The grampus has a bulbous head that hides its beak, and its body is commonly crisscrossed with whitish scars. Possibly solitary and nowhere common, the grampus is distributed throughout the temperate and tropical waters of the world. The most famous grampus of all was Pelorus Jack, who around the turn of the century regularly accompanied ships across Pelorus Sound in New Zealand.

Genus: *Peponocephala*

Peponocephala electra — little blackfish, many-toothed blackfish, melon-head blackfish, melon-headed whale

Found in the tropical Atlantic, Pacific, and Indian oceans.

Genus: *Feresa*

Feresa attenuata — pygmy killer whale

Little is known about this solitary creature, which is rare and restricted to the tropical reaches of the Atlantic, Indian, and Pacific oceans. The animal is known from only a few skeletons and from a specimen kept briefly at Sea Life Park in Hawaii, where it proved a disagreeable customer. The pygmy probably lives its life as a lone predator.

Genus: *Pseudorca*

Pseudorca crassidens — false killer whale

A large dolphin that grows to as much as 18 feet, the false killer whale, despite its name, looks much less like the orca than the pilot whale. It is a slim black dolphin with a prominent dorsal fin, and its smoothly sloping head lacks the bulbous melon of the pilot whale. The false killer is a pelagic species, but schools of the animals, which are recorded into the hundreds and even thousands, do sometimes come close to shore and strand. One such beaching, on October 10, 1946, at Mar del Plata, Argentina, involved 835 whales. The false killer whale appears to prey on large oceanic fish like tuna, mahi-mahi, and bonito, and, although pelagic, it succeeds well in captivity, where it is tractable and quick-witted. The false killer's range extends through the temperate and tropical seas of the world.

Genus: *Globicephala*

Globicephala malaena — longfin pilot whale, blackfish, pothead

This species of pilot whale is found in the temperate portion of the North Atlantic and in the temperate Southern Ocean. Like the other member of this genus, it has a black or dark brown back, a thick and curved dorsal fin, strikingly long flippers, and a bulbous melon that appears above the surface as the animal blows. Pilot whales are social, traveling in pods behind a leader, or pilot, a habit that gives them their name. Pilot whales feed on squid at night, resting by day, and are probably more numerous than daytime sightings might indicate. They show little fear of humans and often approach boats curiously. Pilot whales, like false killers, sometimes strand, but the reason for this behavior is still unknown. Pilot whales do well in captivity and train readily.

Globicephala macrorhynchas — short-finned pilot whale, blackfish, pothead

Similar to the longfin pilot whale, the short-finned species occurs in the tropical and warm temperate waters of the Indian, Atlantic, and Pacific oceans.

Genus: *Orcinus*

Orcinus orca — orca, killer whale

The orca, with its sparlike dorsal fin and brilliant white on black coloring, is hard to mistake at sea. It is found in all oceans, usually along coasts, and is most common in colder, even polar, waters. In the continental United States, killer whales are most numerous in Puget Sound and connected waters, and they are also seen regularly along the southern California coast and about the Channel Islands.

Orcas are social predators that hunt whatever is most available. They adapt well to captivity and have, because of their undeserved reputation for ferocity and the human fear of large predators, become very much the attraction at aquariums and oceanariums.

Genus: *Orcaella*

Orcaella brevirostris — Irrawaddy dolphin, lumbalumba
An inhabitant of both coast and large rivers, this dolphin occurs from the Bay of Bengal to New Guinea and northern Australia.

Genus: *Phocaena*

Phocaena phocaena — harbor porpoise, common porpoise
This small cetacean, which reaches only about six feet and perhaps 160 pounds, frequents coastal waters, estuaries, and bays of both North Atlantic and North Pacific. Its back is a uniform dark color with a small triangular dorsal fin. The harbor porpoise lacks a beak. It tends to be secretive and shy. Long exploited for food, the North Atlantic population is thought endangered.

Phocaena sinus — Gulf of California porpoise, *vaquita, cochito*
Closely related to the harbor porpoise, this species is found only in the upper end of the Gulf of California (Sea of Cortés). It is smaller than the harbor porpoise and more brown than black.

Phocaena dioptrica — spectacled porpoise
This species is restricted to the Atlantic coasts of Uruguay and Argentina, the Falkland Islands, and South Georgia.

Phocaena spinipinnis — black porpoise
A coastal porpoise of the Pacific and Atlantic coasts of the southern end of South America.

Genus: *Neophocaena*

Neophocaena phocaenoides — finless porpoise
The species is subject to taxonomic debate since there may be more than one form. This animal, whether one species or several, ranges in the rivers and along the coast of Asia from Pakistan eastward to Japan, Korea, Java, and Borneo.

Genus: *Phocaenoides*

Phocaenoides dalli — Dall's porpoise, whiteflank porpoise
The Dall's porpoise, with its low dorsal fin, stout body, sloping beakless face, and striking white side patches on a black body, is easy to identify at sea. This species occurs around the Pacific rim from Baja California to Japan and lives in the offshore waters about 20 miles from land. Usually seen in small groups, Dall's porpoises are very fast, hitting sprint speeds in the neighborhood of 30 knots.

FAMILY
Monodontidae — Arctic Dolphins

The two species of Arctic dolphin lack a dorsal fin and beak, and they have a blunt head. This is the most recently evolved cetacean family; the fossil record extends back only about 1,500,000 years.

Genus: *Delphinapterus*

Delphinapterus leuca — beluga, belukha, white whale
Belugas are born gray, but lighten as they age and finally become a beautiful pure white. Belugas are found in the Arctic Ocean and its connected circumpolar waters, with small isolated populations in the Gulf of St. Lawrence and Cook Inlet. In pursuit of their principal food, the Arctic cod, belugas ascend rivers, sometimes for hundreds of miles. These cetaceans are gregarious, gathering at times into immense shoals of thousands of whales. The beluga is hunted for food and leather throughout the Arctic.

Genus: *Monodon*

Monodon monoceros — narwhal, narwhale
The narwhal easily qualifies as the most bizarre cetacean. As an adult the creature has only two teeth, both in the upper jaw, and the left tooth of the male (rarely the female) grows into a long spiraling tusk. The function of the tusk is uncertain. Very likely, the narwhal provided the real-life model for the unicorn of legend. The narwhal is found in the Arctic North Atlantic only, usually in deep water. Native hunters in eastern Canada hunt narwhals for food and for the tusks, which bring a good price.

FAMILY
Physeteridae — Sperm Whales

The sperm whales have their one blowhole off to the left, an upper jaw markedly longer than the lower, and numerous prominent teeth.

Genus: *Physeter*

Physeter macrocephalus — sperm whale, cachalot
The species that supplied Melville with Moby Dick, the sperm whale is the largest odontocete and a distinctive creature. Males run to about 60 feet at full growth; females, around half that. The overall color is dark purple or gray, often heavily wrinkled, and the sperm whale's boxy head may make a distinct bow wave when the animal swims at high speed. The sperm whale lacks a dorsal fin, having instead a series of bumps down the midline of the back. Because of the leftside blowhole, the spout is oblique.

The sperm whale is cosmopolitan, found at one time of the year or other in all deep waters from the equator to the very edge of the polar ice. Bulls range farther north and south than do cows or young. Sperm whales feed principally on squid, but will eat almost any sort of fish that fortune sends their way. The sperm whale is social, living in small pods and schools that periodically gather into great migratory congregations. This species, particularly in the North and South Pacific, is the mainstay of the present whaling industry. Estimates put the world population at about 630,000.

Genus: *Kogia*

Kogia breviceps — pygmy sperm whale

Much smaller than the sperm whale, the pygmy sperm grows to only 9 to 13 feet. It is also much rarer, with only 100 records of the species since its original description in 1871. It has a small dorsal fin and a rather small mouth mounted on the underside of the head something like a shark's. The pygmy sperm lives in the warm temperate and tropical waters of all seas.

Kogia simus — dwarf sperm whale

This species, even smaller than the pygmy sperm, was described only in 1966. Besides anatomical differences in the skull and teeth, the dwarf sperm is distinguished by its size, about 6 feet in the male, and its higher dorsal fin. The dwarf occurs along the coasts of South Africa, India, Ceylon, Japan, Hawaii, California, Baja California, and the eastern United States.

FAMILY
Ziphiidae — Beaked Whales

This family of medium-size (15 to 30 feet) whales is relatively unknown because of its open-sea habitat and its lack of importance to the whaling industry. The beaked whales have a beak, a dorsal fin set well back on the body, and no tail notch. The upper jaw lacks teeth, and the lower has only one or two pairs. The teeth erupt only in adult males, remaining in the gums in females and young.

Genus: *Berardius*

Berardius arnouxi — southern giant bottlenosed whale, Arnoux's beaked whale

This is a large, social animal that typically runs in pods of ten to twenty. The mature whale is powerfully built and black or gray with white belly patches that are sometimes prominent. This species is found in the Southern Ocean.

Berardius bairdi — North Pacific giant bottlenosed whale, Baird's beaked whale

The largest of the beaked whales and the second largest odontocete (surpassed only by the sperm whale), the male of this species sometimes exceeds 40 feet and is occasionally taken by whalers. The Baird's beaked whale differs from the southern form only in size, and some taxonomists believe that the two should be considered one species. This form occurs in the North Pacific from Japan and California north to the Bering Sea. It appears uncommon, but since it is little sought commercially, its numbers may be greater than assumed.

Genus: *Ziphius*

Ziphius cavirostris — goosebeaked whale, Cuvier's beaked whale
A grayish or light brown animal, the goosebeaked whale has less of a beak than the other ziphiids and a narrower and more pointed head. It frequents all oceans but is rare in polar waters, and it feeds largely on squid. The animal is probably numerous, but it travels alone and is hard to spot at sea.

Genus: *Tasmacetus*

Tasmacetus shepherdi
This species is known from only a few specimens found stranded in Argentina, Chile, and New Zealand.

Genus: *Indopacetus*

Indopacetus pacificus — Indo-Pacific beaked whale
Two specimens, one from Australia and the other from Somalia, represent the sum of knowledge of this species.

Genus: *Hyperoodon*

Hyperoodon ampullatus — North Atlantic bottlenosed whale
The sharply rising forehead of this species gives it the bottlenose look responsible for its name. Calves are born dark in color, but the animals whiten as they age, so that old bottlenosed whales may be yellow-white. A social animal found in small groups of from four to twelve, the bottlenosed whale is also migratory in its range from the Davis Strait and Novaya Zemlya to Rhode Island and the English Channel. The bottlenosed whale has been hunted for dog food and for the high-quality oil of its melon, but the animal's ability to dive deep and fast caused so many injuries and deaths that old-time whalemen often refused to seek it from small boats. Until recently, Norway conducted a small catcher-boat fishery for bottlenoses in the European Arctic.

Hyperoodon planifrons — flathead bottlenosed whale
This bottlenose species occurs throughout the Southern Ocean.

Genus: *Mesoplodon*
This could be the least known large genus of cetaceans. Most species have been identified from only a few stranded specimens. Current thinking has it that these whales live in cold and deep water and feed largely on squid and cuttlefish. The lower teeth in the males appear as tusks outside the mouth when the jaws are

closed. They have no function in eating but may have something to do with fights between males or with sex recognition and mating.

Mesoplodon hectori
Recorded in South Africa, the Falklands, New Zealand, and Australia.

Mesoplodon mirus — True's beaked whale
Found in the North Atlantic and about South Africa.

Mesoplodon europaeus — Antillean beaked whale, Gulf Stream beaked whale
Occurs in the western North Atlantic from the Gulf of Mexico to Long Island.

Mesoplodon ginkgodens — ginkgo-toothed whale, Japanese beaked whale
Specimens taken in Ceylon, Japan, Taiwan, and California.

Mesoplodon grayi — scamperdown whale, southern beaked whale
Strandings have provided remains from South Africa, Australia, New Zealand, and Argentina, with one isolated and anomalous instance from Holland.

Mesoplodon carlhubbsi — archbeak whale, Hubbs's beaked whale
Temperate North Pacific. Apparently very rare. Stranded specimen from Drake's Bay on the central California coast was black with a white hump on the head, and its skin bore numerous scratches and scars of unknown origin.

Mesoplodon bowdoini — deepcrest whale
Restricted to Australia, New Zealand, and Kerguelen Island.

Mesoplodon stejnegeri — sabertooth whale, Bering Sea beaked whale, Stejneger's beaked whale
A cold-water whale that ranges north of the archbeak whale, from Japan and Oregon into the Bering Sea.

Mesoplodon bidens — North Sea beaked whale
An inhabitant of the cool temperate North Atlantic from the Bay of Biscay and the tip of Norway east to Massachusetts.

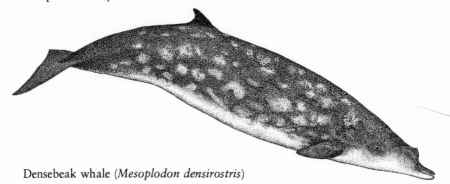

Densebeak whale (*Mesoplodon densirostris*)

Mesoplodon layardi — straptooth whale
  Known from Australia, New Zealand, the Falklands, and South Africa.

Mesoplodon densirostris — densebeak whale, tropical beaked whale, Blainville's beaked whale
  A wide-ranging species of all warm temperate and tropical oceans.

S O U R C E S   A N D
R E A D I N G S

Allen, Gay Wilson. *Melville and His World.* New York: Studio Books. (Viking), 1971. A pleasant book that provides many drawings of America dating from Melville's time and gives a sense of the environment he worked in.

Alpers, Anthony. *Dolphins: The Myth and the Mammal.* Boston: Houghton Mifflin, 1961. Alpers details a variety of mythological and symbolic material about dolphins, tells something about the zoology of the animals, and spins the yarns of Pelorus Jack and Opo.

Andersen, Harald T. (ed.) *The Biology of Marine Mammals.* New York and London: Academic Press, 1969. A collection of scholarly papers on various specialized topics dealing with marine mammals, pinnipeds and sirenians as well as whales.

Anderson, Charles Roberts. *Melville in the South Seas.* New York: Dover, 1966. Originally published 1939. The author traces Melville's travels through the South Pacific. Of interest principally to scholars and devotees of Melville.

Ash, Christopher. *Whaler's Eye.* New York: Macmillan, 1962. The author, an amateur photographer and a professional chemist who worked aboard English factory vessels, provides a pictorial look at modern industrial whaling. Fascinating in a grisly sort of way, with a hard-to-find insider's look at life — and death — on a pelagic whaling ship.

Brower, Kenneth, and Curtsinger, William. *Wake of the Whale.* New York: Friends of the Earth, 1979. An elegant depiction of cetaceans, set in Brower's spare prose and Curtsinger's jewellike photographs.

Bullen, Frank T. *The Cruise of the Cachalot.* New York: Dodd, Mead, 1947. Originally published 1898. A classic account of a South Seas whaling voyage at the end of the Yankee era.

Burton, Robert. *The Life and Death of Whales.* Second edition. New York: Universe Books, 1980. A basic work on whales and whaling, better on whaling than whales.

Caras, Roger A. *Dangerous to Man.* Philadelphia: Chilton, 1964. This definitive work on the dangers posed by wild animals provides the most thorough accounting of attacks by killer and sperm whales.

Carrighar, Sally. *The Twilight Seas: A Blue Whale's Journey.* Illustrated by Peter Parnall. New York: Weybright and Talley, 1975. A fictional tale of the life and death of a blue whale of the Southern Ocean told by one of America's premier nature writers.

Chase, Richard. (ed.) *Melville: A Collection of Critical Essays.* The Twentieth Century Views Series. Englewood Cliffs, N.J.: Prentice-Hall, 1962. This series of essays by literary critics gives an idea of the range, triumph, and folly of current scholarly thinking about Melville's work.

Church, Albert Cook. *Whale Ships and Whaling.* New York: Bonanza Books, 1938. This book contains over 200 photographs of whalers and gear, many dating back into the 1870s and 1880s.

Cousteau, Jacques-Yves, with Philippe Diolé. *Dolphins.* Translated by J. F. Bernard. Garden City: Doubleday, 1975. Better for its photographs than its text; compiled from Cousteau's field research and filming.

——. *The Whale: Mighty Monarch of the Sea.* Translated by J. F. Bernard. Garden City: Doubleday, 1972. Companion volume to the work on dolphins.

Daugherty, Anita E. *Marine Mammals of California.* Second revised edition. Sacramento: Department of Fish and Game, 1972. A useful field guide to California's seals, whales, dolphins, and otters.

Devine, Eleanore, and Clark, Martha. (eds.) *The Dolphin-Smile: Twenty-nine Centuries of Dolphin Lore.* New York: Macmillan, 1967. A collection of dolphin myths, literary allusions, and tales.

Drucker, Philip. *Indians of the Northwest Coast.* New York: McGraw-Hill, 1955. Anthropological handbook published for the American Museum of Natural History. Provides information about whaling among the Indians of the Pacific Northwest and southeastern Alaska.

Fichtelius, Karl-Erik, and Sjölander, Sverre. *Smarter than Man?* Translated by Thomas Teal. New York: Random House, 1971. An essay on the question of whether odontocetes are more intelligent than humans.

Forster, E. M. *Aspects of the Novel.* New York: Harcourt, Brace & World, 1927. This slim book, all of which is a joy, contains Forster's essay about Melville as a prophetic novelist.

Freedman, Russell, and Morriss, James E. *The Brains of Animals and Men.* New York: Holiday House, 1972. A basic work, easy enough for high-school students, about the basics of brain anatomy and the physiology of learning.

Freuchen, Peter, and Salomonsen, Finn. *The Arctic Year.* New York: G. P. Putnam's Sons, 1958. Briefly shows how the cetaceans fit into the yearly cycle of the eastern Arctic.

Friends of the Earth. *The Whale Manual.* San Francisco: Friends of the Earth, 1978. A collection of data and information useful to those campaigning against whaling. Unfortunately the book, first published in England, has a European focus, making it less relevant for Americans.

Gardner, Erle Stanley. *Hunting the Desert Whale: Personal Adventures in Baja California.* New York: Morrow, 1960. The creator of Perry Mason tells how he went to Scammon's Lagoon to look for gray whales when few realized they were there and when travel in Baja California was an ordeal of dust, sun, and *bandidos.*

Griffin, Donald R. *Echoes of Bats and Men.* Anchor Books. Garden City: Doubleday, 1959. An explanation of the biophysics of echolocation by the scientist who discovered it in bats.

Harrison, Richard J., and King, Judith E. *Marine Mammals.* London: Hutchison University Library, 1965. A rather technical book that covers pinnipeds and sirenians as well as cetaceans.

Hawes, Charles Boardman. *Whaling.* New York: Doubleday, 1924. Written close to the end of the Yankee era, this book conveys much of the flavor of that time firsthand.

Henderson, David A. *Men and Whales: At Scammon's Lagoon.* Los Angeles: Dawson's Book Shop, 1972. This is a scholarly work, and therefore topheavy with academic necessities, but it sets the standard for the use of historical data to determine original whale populations.

Hillway, Tyrus. *Herman Melville*. Boston: Twayne Publishers, 1963. A critical work on Melville that combines good sense and a deep knowledge of the author's works.

Hohman, Elmo Paul. *The American Whaleman*. New York: Longmans, Green, 1928. The classic and best study of the economics of the Yankee whaling industry.

Hunter, Robert, and Wyler, Rex. *To Save a Whale: The Voyages of Greenpeace*. San Francisco: Chronicle Books, 1978. The story of Greenpeace, told and photographed by two long-time members. The book is less informative than self-congratulatory.

Jerison, Harry J. *Evolution of the Brain and Intelligence*. New York: Academic Press, 1973. From paleontological information, the author develops a theory about the relationship of brain size to body size and about the evolution of intelligence. A difficult and academic book, but intriguing.

Lawrence, D. H. *Studies in Classic American Literature*. New York: Viking, 1964. Originally published 1923. Chapters 10 and 11 give Lawrence's fascinating readings of *Typee, Omoo,* and *Moby Dick*.

Lilly, John C. *The Center of the Cyclone*. New York: The Julian Press, 1972. Personal chronicle of a journey into the outer reaches of inner space.

——. *Communication between Man and Dolphin*. New York: Crown Publishers, 1978. Lilly's latest work, giving his current thoughts on the topic.

——. *Lilly on Dolphins*. Anchor Books. Garden City: Doubleday, 1975. This one volume contains revised editions of *Man and Dolphin, The Mind of the Dolphin,* "The Dolphin in History," and several scientific papers.

McIntyre, Joan. (ed.) *Mind in the Waters*. New York: Charles Scribner's Sons, and San Francisco: Sierra Club Books, 1974. A collection of writings, poems, and art on, for, and about whales. Although some of the material is poor, more than enough is good to make this a book well worth savoring.

McNulty, Faith. *The Great Whales*. Garden City: Doubleday, 1974. A short work describing the plight and present circumstances of the whales.

Melville, Herman. *Moby Dick*. Afterword by Denham Sutcliffe. New York: New American Library, 1961. Sutcliffe's excellent afterword enhances this edition of the classic novel.

Miller, Tom. *The World of the California Gray Whale*. Santa Ana, Calif.: Baja Trail Publications, 1975. A handy, pocket-sized book that combines a digest of information about whales, particularly the gray, and a field guide to the cetaceans found along the coasts of southern and Baja California.

Mowat, Farley. *A Whale for the Killing*. Atlantic Monthly Press. Boston: Little, Brown, 1972. A frightening book that describes the fate of a finback trapped accidentally and inescapably in a tidal pond. Mowat, a nature writer of considerable accomplishment, uses this sad story to indict human pinheadedness.

Mumford, Lewis. *Herman Melville*. New York: Literary Guild, 1929. One of the earliest and one of the best studies of Melville's life and work.

Murphy, Robert Cushman. *A Dead Whale or a Stove Boat*. Boston: Houghton Mifflin, 1967. Cushman traveled aboard a Yankee whaler after the turn of the century and photographed what he saw, creating a remarkable pictorial legacy.

Nayman, Jacqueline. *Whales, Dolphins and Men*. London: Hamlyn Publishing Group, 1973. A well-written and well-illustrated work on whales and whaling.

Norris, Kenneth. *The Porpoise Watcher*. New York: Norton, 1974. Norris, one of the leading authorities on cetaceans, tells about what he has learned and how he learned it. A pleasant and enlightening book.

——. (ed.) *Whales, Dolphins, and Porpoises*. Berkeley: University of California Press, 1966. A collection of scientific papers largely of scholarly interest.

Orr, Robert T. *Marine Mammals of California*. Berkeley: University of California Press, 1972. A field guide to the seals, cetaceans, and otters of the California coast.

Oswalt, Wendell. *Alaskan Eskimos*. San Francisco: Chandler, 1967. An overall study of the Eskimos, with considerable information about the whaling cultures.

Piggott, Juliet. *Japanese Mythology*. London and New York: Hamlyn Publishing Group, 1969. Source of the story of the whale and the Kamakura Daibatsu in Chapter Six.

Pryor, Karen. *Lads before the Wind*. Introduction by Konrad Lorenz. New York: Harper & Row, 1975. Pryor, a leading student and trainer of dolphins, tells of her experience with the animals and shows how learning by conditioning works.

Rice, Dale W., and Wolman, Allen A. *The Life History and Ecology of the Gray Whale (Eschrichtius robustus)*. Special Publication No. 3. The American Society of Mammalogists. Cast in the conventions of standard zoology, this work is of use only to those interested enough in the gray whale to wade through the academic prose.

Riedman, Sarah F., and Gustafson, Elton T. *Home Is the Sea: For Whales*. Chicago: Rand McNally, 1966. A general introduction to the biology of whales.

Rose, Steven. *The Conscious Brain*. New York: Knopf, 1973. This book is about the brain in humans, not cetaceans, but it describes the current understanding of how the brain came to be and how it works.

Ruhen, Olaf. *Harpoon in My Hand*. London and Melbourne: Angus & Robertson, 1964. An autobiographical account of a whale hunt in Tonga.

Sagan, Carl. *The Cosmic Connection: An Extraterrestrial Perspective*. Produced by Jerome Agel. Garden City: Anchor (Doubleday), 1973. Chapter 24, "Some of My Best Friends Are Dolphins," is worth the reading.

——. *The Dragons of Eden*. New York: Random House, 1977. In this long essay on the brain and intelligence, with an occasional aside about cetaceans, Sagan is always stimulating.

Scammon, Charles Melville. *The Marine Mammals of the Northwestern Coast of North America*. Introduction by Victor Scheffer. New York: Dover, 1968. Originally published 1874. This is Scammon's classic work. Unfortunately, Scheffer's introduction has some errors, such as the wrong name for the ship in which Scammon entered his lagoon. A better source of information about Scammon's life and accomplishments is "Charles Melville Scammon: Whaler Turned Naturalist," by G. Kenneth Mallory, Jr., in *Oceans,* July–August 1976.

Scheffer, Victor B. *The Year of the Whale*. New York: Charles Scribner's Sons, 1969. A recounting of a year in the life of a sperm whale, interspersed with information about whales and whaling.

Schevill, William E. (ed.) *The Whale Problem: A Status Report*. Cambridge: Harvard University Press, 1974. The problem is population, and the papers collected in this volume all address themselves to the question of censusing specific whale stocks. Without intending to, the authors provide a commentary on cetological guesswork.

Seki, Keigo. (ed.) *Folktales of Japan*. Translated by Robert J. Adams. Chicago: University of Chicago Press, 1963. The source for the story of the whale and the sea slug told in Chapter Six.

Slijper, E. J. *Whales*. Translated by A. J. Pomerans. New York: Basic Books, 1962. This is the most comprehensive work on cetacean biology. Unfortunately, much of the work is now out of date, and the book is compromised by Slijper's stubbornly partisan position in favor of the whaling industry.

Small, George L. *The Blue Whale*. New York: Columbia University Press, 1971. Although this book sometimes reads like a Ph.D. dissertation, which it originally was, it provides the most thorough analysis of postwar whaling available, particularly its economic and financial exigencies.

Stackpole, Edouard. *The Sea-Hunters*. New York: Bonanza Books, 1953. A detailed and intriguing study of the New England whalemen from the earliest beginnings of shore whaling in 1635 to the end of the golden age in 1835.

Starbuck, Alexander. *History of the American Whale Fishery: From Its Earliest Inception to the Year 1876.* Two volumes. Preface by Stuart C. Sherman. New York: Argosy-Antiquarian, 1964. Published originally (1878) as a government report to commemorate the centennial of American independence, Starbuck's history is the first and most complete accounting of Yankee whaling, including even an annual list of the vessels leaving and returning to port. Hardly light reading, but an essential research work.

Stone, William Standish. *Idylls of the South Seas.* Honolulu: University of Hawaii Press, 1970. Source for the story of the royal whales of Nuku Hiva in Chapter Six.

Tryckare, Tre. (ed.) *The Whale.* New York: Simon and Schuster, 1968. A well-produced and -illustrated work that provides considerable information about ancient and modern whaling. Unfortunately, the drawings of cetaceans in the list of species are poor, and the "pygmy sperm whale" is not a pygmy sperm whale.

van der Post, Laurens. *The Hunter and the Whale: A Tale of Africa.* New York: Morrow, 1967. This is an elegant novel, set among the whaling fleet based winters in South Africa and told with deep and true feeling.

Walker, Theodore J. *Whale Primer.* Revised edition. San Diego: Cabrillo Historical Association, 1975. This is without doubt the best, most succinct, and most accurate introduction to cetacean biology.

Weaver, Raymond M. *Herman Melville: Mariner and Mystic.* New York: George H. Doran, 1921. This work, more adulatory than critical, started the Melville renaissance. Weaver, unfortunately, assumed that Melville wrote fact thinly disguised as fiction, and he takes the writer at his word, even when he was artistically fanciful. Still, this is a work every Melville fan will enjoy.

Whipple, A. B. C. *The Whalers.* Alexandria, Virginia: Time-Life Books, 1979. The text of this book on the Yankee whalemen is derivative, but the illustrations are many and beautiful.

Williams, Harold. (ed.) *One Whaling Family.* Boston: Houghton Mifflin, 1964. This book combines family chronicles from several whaling members of the same family, giving an interesting firsthand look at the Yankee era.

Wilson, Edward O. *Sociobiology: The New Synthesis.* Cambridge: The Belknap Press of Harvard University Press, 1975. Wilson's comments on John Lilly's work bear thinking about.

Wood, Forrest G. *Marine Mammals and Man: The Navy's Porpoises and Sea Lions.* Washington, D.C.: Robert B. Luce, 1973. Wood looks at dolphins in general and tells about the navy's experience in the marine mammals program.

I N D E X

McNally, Robert, 1946-
    So remorseless a havoc : of
dolphins, whales and men / by Robert
McNally ; illustrations by Pieter
Arend Folkens. -- 1st ed. -- Boston :
Little, Brown, c1981.
    xvi, 268 p. : ill. ; 25 cm.

    Includes index.
    Bibliography: p. 253-259.
    ISBN 0-316-56292-0 : $13.95

1. Cetacea.   2. Whaling.   I. Title.